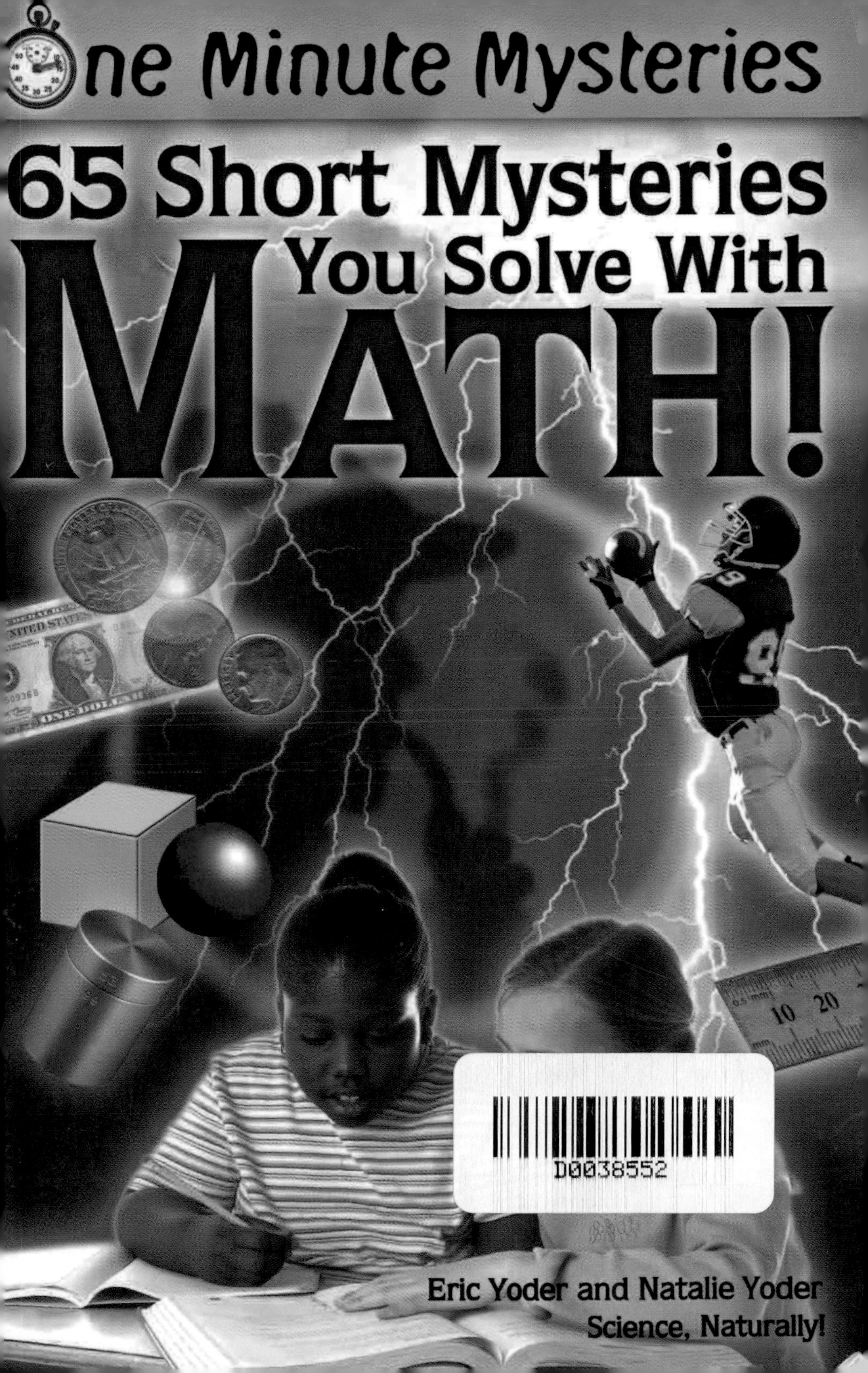
One Minute Mysteries
65 Short Mysteries You Solve With
MATH!
Eric Yoder and Natalie Yoder
Science, Naturally!
D0038552

What the experts are saying about
One Minute Mysteries: 65 Short Mysteries You Solve With Math!

"Keep this book around, and take a moment each day to solve a math mystery puzzle. Yoder and Yoder, a talented father-daughter team, have managed a fine variety of high-energy math stories—from breakfast cereal to a pet iguana, from hula-hoops to the no-longer-planet Pluto. The mysteries are entertaining, and you'll speed along, solving mysteries and sharpening your math skills. What could be more fun?"

—Margaret Kenda, Ph.D., author of *Math Wizardry for Kids*

"What a fantastic way to introduce the application of math to the real world! I've always loved math and, as an architect, I know how applicable it is to the constructed world. But, for my kids, math is not intuitive. *One Minute Mysteries: 65 Short Mysteries You Solve With Math!* engagingly introduces the language of math: fractions, geometry, and algebra. Better than a DVD player for the family van...!"

—Michael E. Burns, AIA, Architect for Capital Projects, George Washington University

"*One Minute Mysteries: 65 Short Mysteries You Solve with Math!* is a fun way to get people thinking about math as a way to find solutions to real problems—not just those you see on a standardized test. These mysteries are the perfect bridge to help the math-phobic embrace the subject as an enjoyable human endeavor rather than a school chore. Get this book for your children and give them a chance to love math!"

—Patrick Farenga, co-author of *Teach Your Own: The John Holt Book of Homeschooling*

"Math often gets a bad rap for being hard and unpleasant. Mysteries, on the other hand, are fun and exciting. The joy of *One Minute Mysteries: 65 Short Mysteries You Solve with Math!* is that each mystery is a math problem in a lively context. Readers get to use their logic and reasoning skills while playing Sherlock Holmes! What a clever way to help young mathematicians hone their sleuthing skills!"

—Ryan McAllister, Ph.D., Biophysics, Georgetown University

"*One Minute Mysteries: 65 Short Mysteries You Solve with Math!* is a new and clever approach to word problems! The mysteries in the book are entertaining, but their real value is in presenting math concepts and quandaries that are approachable, learnable and solvable. A perfect tool to laugh your way to knowledge!"

—Cheryl L. Stadel-Bevans, M.S. Mathematics, Washington, D.C.

One Minute Mysteries:
65 Short Mysteries You Solve With Math!

Eric Yoder and Natalie Yoder

Science, Naturally!®
Washington, DC

1st edition • April 2010 • ISBN: 978-0-9678020-0-8

Ebook edition • April 2010 • ISBN: 978-0-9700106-5-0

Published in the United States by:
Science, Naturally! LLC
725 8th Street, SE
Washington, DC 20003
202-465-4798 / Toll-free: 1-866-SCI-9876 (1-866-724-9876)
Fax: 202-558-2132
Info@ScienceNaturally.com
www.ScienceNaturally.com

Distributed to the book trade in the United States by:
National Book Network
(301) 459-3366 / Toll-free: 800-787-6859 / Fax: 301-429-5746
CustServ@nbnbooks.com / www.nbnbooks.com

Cover, Book Design and Section Illustrations by Andrew Bartelmes, Peekskill, NY

Library of Congress Cataloging-in-Publication Data
Yoder, Eric.
65 short mysteries you solve with math! / Eric Yoder and Natalie Yoder. – 1st ed.
p. cm. – (One minute mysteries)
Includes index
ISBN 978-0-9678020-0-8
1. Word problems (mathematics) – Juvenile literature. I. Yoder, Natalie, 1993- II. Title. III. Sixty-five short mysteries you solve with math!
QA63.Y3. 2010
793.74—dc22

10 9 8 7 6 5 4 3 2 1

Schools, libraries, government and non-profit organizations can receive a bulk discount for quantity orders. Please contact us at the address above or email us at Info@ScienceNaturally.com.

Printed in the United States of America

The characters in this book are a product of the authors' imaginations and are not real people, living or dead.

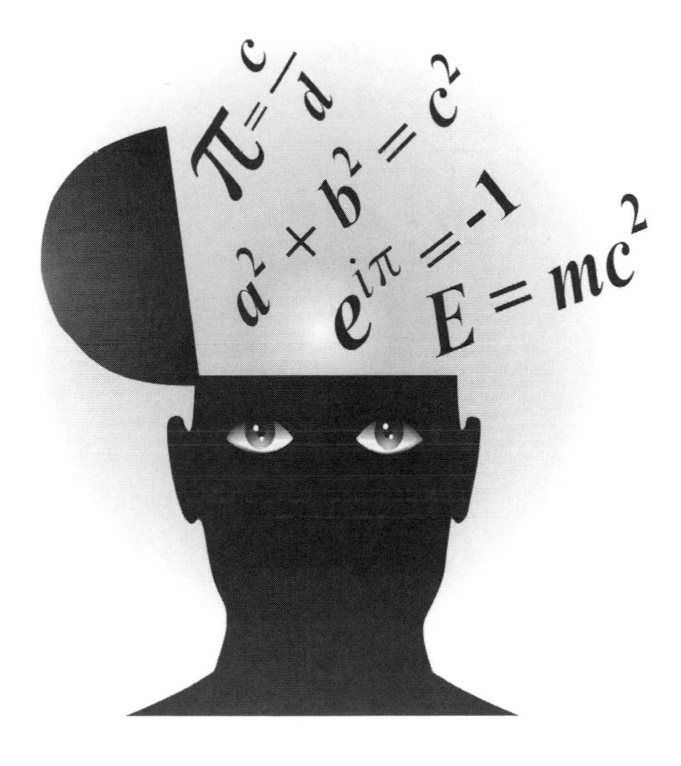

Table of Contents

Why I Wrote This Book–

by Eric Yoder

The words math and science often are paired together, especially in the context of education. So a math mysteries book seemed a natural sequel to *One Minute Mysteries: 65 Short Mysteries You Solve with Science!* The purpose of this book is the same as that of the first—to show real-world applications of academic subject matter, using mystery as the vehicle.

Our goal was not to test how well readers do calculations or remember formulas, although math by necessity involves some of that. Rather, by blending academic subject matter with some light-hearted narratives, we have tried to make math more accessible and enjoyable not only for children, but also for readers of all ages.

It has been gratifying to hear that so many parents, children, teachers, home schoolers and other readers enjoyed the science book for both the mystery and the learning aspects. We hope this book shows that math—often math that you can do in your head—can similarly help solve the everyday mysteries of life.

—Eric

Why I Wrote This Book–

by Natalie Yoder

Math is the kind of thing that you think you're never going to use, but that's not true. You have to use math lots of times in real-life situations, just like in this book.

Many of these mysteries started with the idea of an everyday problem that could be solved using math. Sometimes, my Dad and I would think up a great idea for a story, and we would quickly write it down. I could write them pretty fast, so I wrote a bunch by hand or on the computer and gave them to my Dad to edit. He wrote some and I edited them, and we wrote others together.

When you're faced with a situation that you can't figure out, math can sometimes help you. Writing these stories tested how well we could think of a mystery, and reading them will test how well you can solve them using math. I hope you like them!

—Natalie

Math at Home

1 Heavy Toll

"A speeding ticket? What?" Suzy's father said as he opened the day's mail.

"What's the matter, Daddy?" Suzy asked.

"Well, Suzy, this ticket says that we were speeding on the toll road we took when we were driving back from the state science fair last weekend," he explained.

As drivers entered the road they got a receipt showing the time and exit number. The exit numbers were also mileage markers. When they got off the road, drivers had to pay different amounts depending on how far they went.

"Are you sure they're right?" Suzy asked. "What does it say?"

"Well, it says that we got on at exit 64 at 12:13 p.m., then got off the road at exit 148 at 1:33 p.m.," he said. "And it says the speed limit was 55 miles an hour—I thought it was 65. How can they know if we were speeding?" he asked. "I didn't see any police cars."

"It's too bad, but they're right," Suzy said.

"How do you know?" he asked.

"If we got on the road at 12:13 and got off at 1:33, that means we were on the road for one hour and 20 minutes, or 80 minutes," Suzy explained. "Since the exit numbers are mileage markers, the distance between exits 64 and 148 is 84 miles—148 minus 64. That means we went 84 miles in 80 minutes—that's more than one mile per minute, which is more than 60 miles per hour. So we were speeding, since the speed limit was 55 miles per hour."

"To figure it out exactly," she added, "84 miles divided by 80 minutes makes 1.05 miles per minute. Multiplying 1.05 miles per minute by 60 minutes in an hour to get miles per hour means we averaged 63 miles per hour."

"Well, we were going less than that for some of the time," her father said.

"Yes, but to average 63 miles an hour, we must have been going faster than that at other times," she said. "I hope that ticket isn't too expensive."

2

Role of the Dice

"Five-minute warning, kids!" came their father's voice from the back yard.

He was grilling dinner, and he meant it was time for the table to be set. That was one of the three chores that Kimberly, Quentin and Brian split each evening. The other chores were cleaning up after dinner and taking out the recycling and the trash. The chores were about equal, but, like many evenings, no one wanted to go first.

Kimberly, who was seven years old, was playing Quentin, who was nine, at backgammon on the screened-in porch where they ate supper during the summer. Eleven-year-old Brian was watching the game.

"Whose turn is it to set the table?" Kimberly asked.

Quentin and Brian shrugged. They didn't remember either.

"How about we toss a pair of dice for it?" Quentin suggested. "Whoever's age comes up first sets the table, and whoever's age comes up second clears it."

"That seems fair," Kimberly said.

"No, it's not," Brian said.

"Sure it is," Quentin said. "You can't control how the dice will come out, so each of us has an equal chance of our age coming up. What can be fairer than that?"

"It's true, you can't control how the dice will come out," Brian said, "but that doesn't mean our ages have an equal chance of coming up."

"Why wouldn't they?" Kimberly asked.

"When you roll two dice, the combined numbers can fall between 2 and 12," Brian said. "There's only one way to get a 2—a 1 on both dice—and only one way to get a 12—a 6 on both. There are two ways to get a 3 or an 11. To get a 3, you can have a 1 on the first die and a 2 on the second, or a 2 on the first die and a 1 on the second. To get an 11, you can have a 6 on the first die and a 5 on the second, or a 5 on the first die and a 6 on the second."

"The pattern goes on that way," Brian said. "There are three ways to get either a 4 or a 10, four ways to get a 5 or a 9, five ways to get a 6 or an 8, and six ways to get a 7. That means that when you roll two dice, the number most likely to come up is 7. Since Kimberly is seven years old, she's the most likely one to have to set the table."

"It won't necessarily happen that way, though," Quentin said. "Any number from 2 through 12 still can come up."

"True," Brian said. "But we're talking about probability here. On any roll of two dice, the number most probable to come up is Kimberly's seven. And your age of nine, Quentin, is more probable to come up than my age of eleven."

3 Pancake Mix-up

"Mooommm!" Meg yelled from the kitchen. "Can you please come down here?"

Meg's family and two other families had rented a house at a ski resort for a long weekend. Each family was going to cook and clean up for one of the three days. It was the morning of Meg's family's day.

While Meg's mother finished getting dressed, Meg went into the kitchen and started preparing the pancake mix. They had brought individual-sized serving packages of mix. They also had several boxes of cereal and bread to make toast, but everyone had said they wanted pancakes.

"I'll be there in a minute, Meg. What's the problem?" her mother called.

"I have everything ready to make the pancakes. But each of these packages needs two-thirds of a cup of milk, and there's no two-thirds measuring cup in this kitchen," Meg called. "All they have is a three-fourths measuring cup. Can I just estimate?"

"Not if you want the pancakes to be any good," her mother replied.

"Never mind," Meg said a moment later. "I have the solution."

"What did you do?" her mother asked as she walked into the kitchen.

"I did some math. It's a question of least common multiples," Meg told her mother. "First, I figured out how many times you'd have to fill each kind of measure to reach a whole number. With the three-fourths measuring cup, to reach a whole number you'd need to use the measure four times. Four times three-fourths is twelve-fourths, which reduces to three. So filling that measure four times gives us three cups of milk.

"Each package of mix required two-thirds of a cup of milk. If we had a two-thirds measuring cup, you would need to fill it three times to get a whole number. Three times two-thirds is six-thirds, which reduces to two. So, filling a two-third measuring cup three times would give us two cups of milk," she continued.

"All I had to do then was find the least common multiple of three and two—the smallest number that is a multiple of both. That's six. Since I would need to fill the three-fourths measuring cup four times to get three cups, I would need to fill it twice that many times, eight times, to get six cups. I did that and put the milk in the bowl. And since three fillings of a two-thirds measuring cup would give us two cups, to get six cups I would need three times that many, or nine, to get the right amount of mix. So I added nine packages of the mix. I hope everyone's hungry!"

4 In Hot Water

"Aida, it's time for us to clean out the fish tank," Zoe said.

The sisters shared a room that had two catfish and a Betta fish named Sparkle in a five-gallon tank. They put the fish in a small jar, took out the old water, and cleaned the tank and gravel.

They had a two-gallon bucket that they used to put fresh water into the tank.

"Oh no, I just remembered," Aida said, looking at the thermometer in the tank after they had put in two buckets' worth. "The temperature's supposed to be 72 degrees Fahrenheit, and the water is only 65 degrees."

"We could wait until the water comes to room temperature—that's about 72 degrees," Zoe said.

"I hate to leave the fish in that little jar for that long," Aida replied. "Maybe we could just make the last bucket of water hotter. Since there are four gallons of water in the tank at 65 degrees, and we want the temperature to be 72 degrees, we have to raise the temperature of each gallon by 7 degrees. So the next bucket we put in should be the right temperature above 72 degrees to do that. Since we have a two-gallon bucket, we should fill the bucket with water that's 14 degrees above 72, or 86 degrees. That way each gallon coming from the bucket would warm up two of the gallons already in the tank by 7 degrees, and all the water would be 72 degrees."

"I don't think that will work," Zoe said.

"Why not?" Aida asked.

"You're forgetting one thing—there are already four gallons of water in the five-gallon tank," Zoe said. "So there's only room for one more gallon, not two. We'd have to make that one gallon 28 degrees above 72 to bring the other four gallons up by 7 degrees each. That means the last gallon has to be 100 degrees for the different temperatures of the water to reach a balance at 72 degrees."

5 Flooring Them

Lily and Robert agreed with their parents that it was time to replace the worn-out kitchen floor. It was cracked, stained, and impossible to clean.

But picking new tile was a different matter. Their parents had brought home half a dozen samples from the tile store and laid them out around the kitchen. Although the floor was 10 feet by 12 feet and the samples were only one square foot, they were big enough for everyone to get a feel for how they would look.

In the end it came down to a choice of two kinds of tiles. The imitation granite tiles came in boxes of 25 for $100 a box. The imitation marble tiles came in boxes of 50 for $150 a box. In both cases, they had to buy an entire box, and unused tiles couldn't be returned for a refund.

Everyone in the family agreed they liked the two kinds equally—the decision was just a matter of which was cheaper.

"Let's buy the granite ones," Robert said. "We'll have a lot less left over."

"But aren't we trying to save money?" Lily asked.

"That's what I meant," Robert said.

"Well, that's what I meant, too, and I think we should go with the marble ones," she said.

As usual, the argument ended with an appeal to their parents.

"Okay, each of you tell us why you think one tile will save more money than the other," their father said.

Robert said, "Well, the area that needs to be covered is 10 feet by 12 feet. That's 120 square feet—10 times 12. If we buy the granite tiles, which come in boxes of 25, we'd need five boxes—five times 25 is 125. We'd have 5 tiles left over—125 minus 120. Since it costs $100 for a box of 25, each tile costs $4, or 100 divided by 25. With 5 left over, we're wasting $20, 5 leftover tiles times $4 apiece."

He continued, "With the marble tiles, there's 50 in a box, so we'd need three boxes to cover the 120 square feet, three times 50 is 150. We'd have 30 left over, or 150 tiles minus 120 square feet. Since those boxes cost $150 each, each tile costs $3, or $150 divided by 50. So we'd be wasting $90, 30 leftover tiles times $3 apiece."

"True," Lily said, "but it's a question of how much we're spending in total. The five boxes of granite tiles would cost $500, or five times $100. The three boxes of marble tiles would cost only $450, three times $150. So we'd be saving $50 by buying the marble tiles, even though we would have more tiles left over."

"Lily's idea actually would save us money," their mother said. "Let's go with the marble tiles."

6 Compounding His Interest

"But Grandpa, college is a million years away!" Damien said.

Damien's family was having a party to celebrate his 8th grade graduation. He would be going to high school in the fall.

"I'm sure it seems like a long time to you," his grandfather said. "But it's time we started making sure you'll have enough money for college. So, here's what we're going to do. We've opened a bank account and put $1,000 in it for you."

"A thousand dollars!" Damien exclaimed.

"College is expensive," his grandmother said. "It's because we went to college that we can afford to do this. And we intend to do this each summer for the next four years, too. Plus, we're going to increase what we give you by 10 percent each year, because college is getting more expensive all the time."

"I don't know what to say, except thanks so much," Damien said. "Let's see, by the time I start college, I'll have, um, $5,400."

"Will you?" his grandfather asked.

“I see what you mean,” Damien said. “It will be more. At first I thought that by adding 10 percent next year and for the three following years, you were talking about 10 percent on the $1,000 each year. So that would have been $5,400—this year’s $1,000 plus $1,100 for four more years.

“But you said you were going to increase what you gave by 10 percent each year, which means compound interest. So, next year you would give me $1,100—10 percent more than $1,000. In the third year, it would be $1,210—10 percent more than $1,100.”

His grandfather took out a sheet of paper to do the rest of the calculations. “In the fourth year, it would be $1,331, and in the fifth year, the year you finish high school, it would be $1,464. So it would come to $6,105 in all,” he said. “That shows you the value of compound interest. You get interest each year not only on the original money but also on the interest you got in earlier years.”

“It’s very generous of you. Thank you so much!” Damien said.

7 Setting the Date

One night, Elijah and Kevin were watching the debate for the upcoming Presidential election. They were in Kevin's family room, taking notes about the topics that were being discussed and the main points each candidate was making. There was going to be a quiz on it in social studies class the next morning.

Kevin's younger brother, John, was sitting at the computer in the corner, working on the invitations for his birthday party, which was going to be on January 6. He always had his party on the actual day of his birthday, even if it was a school day. He was getting excited about it already, even though it was more than two months away.

Kevin had helped him get started on the computer program but had to move over to the TV when the debate came on.

John read aloud what he had written so far: "You are invited to a party for John's birthday on . . . something, January 6. Kevin, what day of the week is my birthday going to be? We don't have a calendar for next year yet."

"How should I know?" Kevin said. "I'm trying to watch this debate."

"Do you remember what day of the week your birthday was on earlier this year?" Elijah asked John.

"Sunday," John replied. "I remember we watched a pro football game on TV during the party."

"Well, then shouldn't it be obvious which day of the week your next birthday will be?" Elijah asked.

"Oh, right," Kevin said. "Each year the day of the week that a date falls on is one day later than the previous year. There are 365 days in a year, but 52 weeks at 7 days each is 364 days—52 times 7. So there's one extra day, which pushes back the day of the week by one. So a date that falls on a Sunday one year will fall on a Monday the next year. So put Monday, January 6 on your invitation," he said to John.

"Actually, make it Tuesday, January 6," Elijah said. "The rule Kevin said is true except after leap years. In a leap year, there are 366 days, or 2 extra days, so a date the following year will fall 2 days later in the week. And this is a leap year, since Presidential elections always happen in leap years. That means your birthday will fall 2 days later, on a Tuesday."

8 Corralling the Problem

Nicole's little sister Valerie loved play sets. She had several of them set up in their family room, which her family called Valerie Land. There were houses, a petting zoo, a bakery and lots of farm animals. She especially loved horses.

At her birthday party, she got several sets of horses, including one set that had 40 plastic fence pieces, each one inch long, that could be snapped together to make a straight line or right angles. After the guests went home, Valerie started to play with the fence.

"I hope this corral is big enough to hold all of my horses," she told Nicole.

Nicole knew that would be a challenge because Valerie had a lot of horses.

Valerie arranged the fence pieces into a rectangle that was much longer on two of the sides than on the other two, but she couldn't fit all the horse figures inside of it. "We don't have enough fence pieces," she said. "We need to buy more."

"Before we do that, let me try to help," Nicole said, starting to rearrange the pieces.

"What difference will that make?" Valerie asked. "We have the same number of pieces no matter what shape we make."

"Let's arrange the fence pieces into a square," Nicole said. "We have 40 pieces, so a square would have 10 pieces on each side. The area of a square is the width times the length, 10 times 10, or 100 square inches.

"Let's say you make a rectangle that's as close as you can get to a square. That would be 9 pieces in one direction and 11 in the other. That gives an area of 99 square inches—9 times 11. That's not much of a difference, but 99 is smaller than 100. Or, a rectangle that's 8 pieces one way and 12 the other would have an area of 96 square inches—8 times 12. If you go all the way to a rectangle that's 1 piece one way and 19 pieces the other way, you have an area of only 19 square inches—one times 19. So a square is the shape that will enclose the most space. Let's see if all of your horses can fit into a square."

They did fit. As Nicole watched, she thought to herself, I didn't want to confuse her, but a square is actually a type of rectangle, since a rectangle is a four-sided object with all straight lines, four right angles, and opposite sides of equal length. A square is a kind of rectangle where all four sides are the same length. I used the word rectangle in the sense that people usually think about it, where one pair of sides is longer than the other pair.

9 It's a Gas

"No kidding, we're getting a new car?" Olivia said excitedly as she came into the living room, where her parents and her brother Daniel had spread out information about new cars. Their old one had lasted 10 years, but now it was time for a new one.

Daniel had his eye on a sports car that would cost $27,000. Their mother liked a van that would cost $25,000, while their father was leaning toward an SUV that would cost $29,000. Olivia liked a hybrid that would cost $28,000, but they all agreed that they would be happy with any of them.

"In that case, shouldn't we just get the least expensive one?" Daniel asked.

"It's not just the purchase price, it's also how much it costs to run it," their mother said. "The maintenance costs seem about the same. The main thing is gas, which costs about $2 a gallon now. The sports car and the van each get 25 miles to the gallon, the SUV gets 20, and the hybrid gets 40."

Their father added, "We drive an average of 10,000 miles a year. Probably we'd keep the new car as long we kept the car we have now."

Olivia thought for a minute. "Then math alone won't make the decision for us," she said.

"What do you mean?" their mother asked.

"Well, to find out how much we'd spend on gas with each one, we need to know how many total miles we will drive it. If we expect to keep the car for 10 years and drive 10,000 miles a year on average, that's 100,000 miles—10,000 times 10," Olivia said. "To figure out how much gas we would use, you divide those 100,000 miles by the miles per gallon each car gets."

Daniel said, "The sports car and the van that get 25 miles per gallon would use 4,000 gallons each over 100,000 miles. The SUV that gets 20 miles per gallon would use 5,000 gallons, and the hybrid that gets 40 miles per gallon would half of that, 2,500 gallons."

"At $2 a gallon," Olivia said, "the sports car and van each would use $8,000 of gas, the SUV $10,000 and the hybrid $5,000."

"Finally, add the cost of the gas to the cost of the car," Daniel said. "For the sports car, that's $27,000 plus $8,000—$35,000. For the van, that's $25,000 plus $8,000—$33,000. For the SUV, that's $29,000 plus $10,000—$39,000. And for the hybrid, that's $28,000 plus $5,000—$33,000."

"That means the van and the hybrid would cost the least but the same—$33,000 in total," Olivia said. "That's what I meant when I said that math alone won't make the decision for us."

"I guess we don't really need the extra space in the van, so I'd say we should buy the hybrid," their mother said.

"I agree," their father said. "Burning less gas is better for the environment."

10 Cover Up

As a birthday present to her little sister Laura, Miranda had promised to paint the inside of the family playhouse for her.

Years before, their father had painted the walls and floor pink, Miranda's favorite color. But since Laura was the one who mainly used it now, and her favorite color was blue, she wanted the pink covered up.

Miranda measured the inside of the playhouse. The two longer sides were 10 feet long and 6 feet high, and the ends were 6 feet long and 6 feet high. Above that was the inside of the roof, which didn't need to be painted. Her father warned her that covering up the pink would require two coats of paint.

Later at the hardware store, Laura chose a shade of blue that she liked.

"Okay, here's a can that says it will cover 520 square feet," Miranda said. "Each longer side of the playhouse is 60 square feet—10 times 6—so together they would be twice that, or 120 square feet. The ends are 36 square feet each—6 times 6—so together they would be twice that, or 72 square feet. And 120 plus 72 is 192 square feet. Painting that twice means I need to cover 384 square feet in total—two times 192. So a can that covers 520 square feet will be enough."

Since she was paying for it out of her own money, Miranda didn't want to buy too much.

"That's enough to cover the walls, but don't forget you have to paint the floor, too," her father said.

"Oops! I didn't measure the floor," Miranda said.

"Should we drive back home to measure it?" Laura asked. "Or should you just buy an extra can of paint to be sure you have enough?"

"Neither," Miranda said. "Since we know the two longer sides of the playhouse are 10 feet long and the ends are 6 feet long, the floor must be a 6 foot by 10 foot rectangle, meaning its area is 60 square feet. Painting that twice means I have to cover another 120 square feet. So I need to cover 504 square feet—384 plus 120—in total. That means one can will still be enough."

Cereal Numbers

Ron and Lauren's father had made a New Year's resolution in 2009 to be healthier. Part of his program, along with exercising, was eating only cereal for breakfast, instead of things like bacon and eggs.

It had taken him a while to find a kind of cereal he liked, but finally, on the first day of February, he settled on one. It happened to be Ron and Lauren's favorite too.

He even had his own special box of it, which he labeled with a marker: "Dad's Box—Not for Kids." The box held 700 grams of cereal, and by measuring out a cup a day, he'd made the cereal last exactly one month. So he decided that he'd start a new box on the first of each month.

One day in late March the family sat down to breakfast together. Their father looked into his box and frowned.

"I don't think I'll have enough to make it through the month," he said.

He looked at Ron and Lauren. "This reminds me of Goldilocks and the Three Bears," he said. "Someone's been eating my cereal," he said in a deep voice.

"No we haven't, Dad," Ron and Lauren said together.

"Then what happened to my cereal?" he asked.

"The first box lasted all of February, which has 28 days this year-every four years it has 29," Lauren said.

"But March has 31 days," Ron added. "So to make this month's 700-gram box last 31 days, you should have been taking out a little less each day."

Lauren did some quick division on a sheet of paper. "To make 700 grams last 28 days, the measuring cup must hold 25 grams of this cereal," Lauren said.

Ron also did some division. "You should have been taking 22.58 grams this month, to be exact. But don't worry, we'll give you some from our box."

12 Toss-Up

"These cookies must be for me," Dylan said.

"No, they must be for me," Isaac said.

Dylan and Isaac's travel basketball team had stopped for dinner at Dylan's house after a game. Dylan's mother had made a fantastic dinner, and everyone, except Dylan and Isaac, was too full for the cookies Dylan's father had baked.

Isaac said, "Let's toss for them. Cookie by cookie." He got out a quarter. "I'll toss, you call," he said.

"Heads," Dylan said.

It came up tails. Isaac ate a cookie.

"Heads," Dylan called on the second toss.

Tails again. Isaac ate another cookie.

"Heads," Dylan called again.

It came up tails again. Isaac ate a third cookie.

By now the other boys were snickering. "Heads again," Dylan said.

Tails once more.

"Maybe you should start calling tails," Isaac suggested. "I'm getting pretty full, eating all these cookies."

"No, I'll stick with heads," Dylan said. "I mean, what are the odds that it will come up tails again?"

"I can tell you the odds exactly," Isaac said.

Dylan looked surprised. "How did you figure it out that fast?" he asked.

"The chances are even that on any coin toss, either heads or tails will come up," Isaac said. "It doesn't matter what happened on any previous tosses. The odds are still 50-50 that it will come up one way or the other the next time."

"By the way," Isaac added, "it's easy to figure out the odds of a coin toss coming up one way or the other five times in a row. You double the result each time. The chance of a coin coming up one way or the other is one in two the first toss. The chance of it coming up a certain way each time is one in four for two tosses, one in eight for three tosses, one in 16 for four tosses and one in 32 for five tosses. But that's just the odds against that happening in general. On any given toss of the coin, the odds are always 50-50."

13 Seeing the Light

"Anna, could you read over Noah's story for him?" their father asked.

"Okay, Dad," she said.

Anna's younger brother Noah was in third grade and was just starting to write stories. Since Noah liked space and rockets, she wasn't surprised when he handed her a story about a spaceship.

It read:

Once upon a time, there were three astronauts who took a trip to Pluto. They were sad because Pluto isn't a planet anymore. Since Pluto is far away, it took them a long time to get there, two whole light years. When they got there, they found people who asked them to make Pluto a planet again. They stayed there a year, and it took them two more light years to get back. Everyone wondered why they had been gone six whole years, but when the astronauts told them about the nice people on Pluto, they decided to make it a planet again. The End.

"What a great story," Anna said. "Can I edit it just a little for you?"

"What do you want to change?" Noah asked.

“A light year is a measure of distance, not time,” she said. “It’s the distance light travels in one year in a vacuum such as space. That’s nearly 5.9 trillion miles, or in metric, nearly 9.5 trillion kilometers. So just have them traveling two years, not two light years, on each trip.”

“And by the way,” she added. “Once you fix that, either have them stay on Pluto for two years instead of one, or change it so that when they come back they have been gone five years, not six.”

14 All Wound Up

"No way, a real wind-up watch? One that ticks and everything?" Ian asked.

"I think I saw one of those in an old movie once," Wyatt said.

Hector was showing off a watch that had been in his family for many years and was a family treasure.

"My grandfather gave it to me today," Hector said. "It was right at noon. He showed me how to set the hands and wound it one turn of this little wheel on the side to start it."

"Are you sure it works, though?" Wyatt asked. For the first time, Hector noticed that the watch was not running. Its hands showed three o'clock, and it was now four o'clock.

"Well, I guess I have to wind it again," Hector said.

"Isn't that going to be a lot of trouble?" Ian asked. "I mean, winding it again and again every day."

"I don't think it will be that often," Hector said. He started winding the watch, counting sixteen turns until it was fully wound. "In fact, I can tell you exactly when it will need to be wound again."

"When is that?" Wyatt asked.

“Four o’clock, the day after tomorrow,” Hector answered.

“How do you figure that?” Ian asked.

“Well, my grandfather started it by winding it one turn, and it ran for three hours,” Hector said. “I just now wound it all the way, sixteen turns, which means that when you wind it fully it will run for 48 hours before stopping—sixteen times three. Winding it every other day isn’t too much trouble, especially since it means so much to me.”

15 Getting the Point

Riley heard Sofia, Cameron and Hunter coming home. It was a few days before school was starting and their mother had taken them shopping for supplies. Riley had done her back-to-school shopping a few days earlier.

Their mother had said she would pay for the basic supplies, but if they wanted anything special, they would have to use their own money from their allowances.

As the younger children emptied their bags on the family room floor, Riley noticed a box of 100 great-looking pens.

"I wish I'd bought some of these for myself," she said. "Did each of you buy one of these boxes?"

"None of us had enough money left to buy this big box by ourselves," Hunter said.

"Mom said it would be better to buy one big box rather than three smaller ones," Cameron said. "So we pooled the money we had left."

"The trouble is, now we can't figure out how many pens each of us should get," Sofia said. "I know I should get the most, because I put in the most, $8. Hunter only had $2 and Cameron only had $1. Can you help us?"

"It's just a question of algebra," Riley answered. "Cameron donated the least, $1, so we'll call her donation 1x. Hunter, you donated $2, so you donated 2x. Sofia donated $8, so she donated 8x. That means the total donation was 1x plus 2x plus 8x—11x. Since there are 100 pens in the box, that gives us a formula of 11x equals 100. To solve, divide both sides by 11, meaning that x equals 9, with one pen left over. So Cameron gets 9 of the pens, Hunter twice as many, 18, and Sofia gets eight times as many, 72. A total of 99."

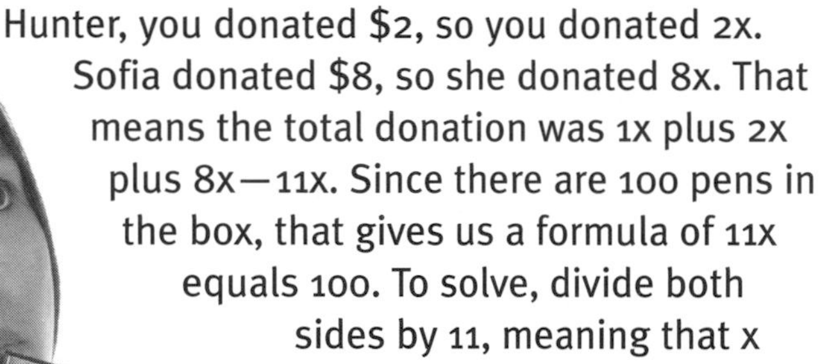

"And the one left over," Riley added, taking a pen and turning to go to her room, "is for me for figuring that out."

Math Outside

16 Tall Tale

Challenge Day was one of the highlights of the week at camp. The campers were sent off on all kinds of odd errands, such as finding animal fur, certain kinds of leaves, nuts and other bits of nature.

Dominic and Vincent had ended up with what they were sure was the toughest assignment: figuring out the exact height of the lone tree in the center of the field.

They almost had to laugh when they were given only two tools to do it: a yardstick and a large ball of string.

"This is impossible," Dominic said, squinting up at the top of the tree. It was a sunny day.

"It can't be impossible," Vincent said. "The counselor said other guys have done it just with the same things they gave us."

They thought for a while.

"Well, I have an idea," Dominic said. "But it's not going to be easy. One of us could hold the ball of string while the other one ties the end to his belt and climbs the tree. We could probably get close enough to the top to estimate how much was left, and then we could add that to the length of string from there to the ground."

"I don't think we'd like the result of that," Vincent said.

"Why not?" asked Dominic.

"Even if the climber didn't fall out of the tree, we still wouldn't have an exact answer," Vincent said. "How about if we put one end of the yardstick on the ground, hold the stick straight up, mark the end of the shadow and measure the length of the shadow. Then, we'll run the string from the base of the tree to the tip of the tree's shadow, and use the yardstick to measure how much string was used."

"What good will that do?" Dominic asked.

"It's a matter of ratios," Vincent said. "The ratio of the yardstick's height to its shadow will be the same as the ratio of the tree's height to its shadow. Let's say the yardstick makes a 2-foot long shadow. That would be a ratio of 2 feet of shadow for every 3 feet in height. So, for every 2 feet of the tree's shadow, the tree would be 3 feet high. Say the tree's shadow is 40 feet. That would mean the tree is 60 feet high."

"What if we don't get such nice round numbers?" Dominic asked.

"It's just a matter of doing the math, we'll still get the answer," Vincent said. "And we won't have to climb the tree."

17 Raking their Brains

"Hannah! . . . Would you please come out here already? Your sister, father and I have been waiting for you for almost 10 minutes!"

"Coming, Mom!" I hate the middle of November, when all the leaves have fallen and we have to rake them, Hannah thought to herself.

After half an hour of raking leaves, Hannah had blisters all over her hands, but they were almost done. Only one more pile was left. The trouble was that the pile was in the middle of two rows of cars and vans, and they had to move it out to where the leaf collecting truck could get it.

"Hey, Juliet," Hannah called to her sister. "I'll race you, and whoever can rake their half of the pile to the other side of the cars first wins."

"Okay, but the loser has to clean up the rakes!"

They divided the pile into equal halves. Juliet looked at the two rows of vehicles on either side of her. The left row had three vans, parked 4 feet apart from each other. The right had four cars parked the same distance apart.

"Dad, how long is a car and a van?" Juliet asked.

"A car is about 15 feet, a van about 20," he said.

"We should each take one row to rake around," Hannah said.

"I'll take the one on the left," Juliet said.

"Okay."

"On your mark, get set . . . GO!" Juliet screamed.

After a few minutes of speed raking Juliet proudly remarked that she was done.

"You cheater!" Hannah said.

"I did not cheat, Hannah. Look, I'll explain . . ."

Juliet said, "When you told me to pick a side, I did some estimating. On the left side, three vans, about 20 feet each. On the right side, four cars, about 15 feet each."

"So that's equal," Hannah said. "Three vans times 20 feet each is 60 feet on the left, and four cars times 15 feet each is 60 feet on the right."

"You're forgetting the distance between them—4 feet each," Juliet said. "Three vans in a line means there were two gaps between them—a total of 8 feet. But the four cars have three gaps between them—a total of 12 feet. Even though it was only about 4 feet less on the left side, I thought it would be enough to help me win. Each time I went back for more leaves I was going 4 feet less and I had to rake them 4 feet less."

After Hannah cleaned the rakes, they had too much fun jumping in the leaves to talk about it anymore.

18 A Measured Response

"Mix contents of package with two gallons of water," Steve read from a package of plant fertilizer.

Steve and Les were counselors-in-training at a summer camp. Their job at the moment was to go down by the front entrance near the creek to fertilize and water the flowerbed beneath the entrance sign. The first campers of the summer would be coming the next week and the camp director wanted everything looking its best.

They walked the half-mile down the road to the entrance. Because it was so hot, they sipped on sodas from the dining hall—a regular sized can for Steve, and a regular sized plastic bottle for Les. Les was carrying an empty watering can. He examined it.

"Um, there are no markings on this thing," Les said. "Maybe we could just estimate."

Steve read again from the package: "Caution: Mix contents exactly."

"So much for estimating," Steve said. "I know what to do. I'll finish my soda, then go down to the creek and measure out two gallons of water using my soda can."

"I know how to do it faster," Les said. "Two fifths faster, to be exact."

"How?"

"We'll use my bottle instead of your can," Les said. "Soda cans hold 12 ounces. Plastic bottles hold 20 ounces. So your can is twelve-twentieths the size of my bottle, which reduces to three-fifths. For each 3 ounces you'd get by filling your can in the water, I'd get 5 by using my bottle. In other words, each time I'd dip, I'd get two fifths more water than you.

"Better yet, let's use both," Steve said. "We'd only have to fill them eight times. Combined, they hold 32 ounces, which is a quart. And since there are four quarts in a gallon, to get two gallons we need eight quarts."

19 Lawn Ranger

"Dad, we really need a new lawn mower," Murphy said.

Of all people, Murphy would know. Since he'd turned 12, it had been his job to mow the lawn every Saturday morning. The family rule was that it had to be done before any fun activities.

Murphy had done it so often that he even knew the exact number of times he had to go up and down the yard—20 times in each direction.

His father looked at the old lawnmower. It was hard to start, ran rough, and left the grass in clumps. "You're right. Let's go to the store and see what they have," he said.

At the store, Murphy's father pointed one out. "Here's the newest version of the model we have, with the same cutting blade, 24 inches across," he said. "And here's one with a 30-inch blade."

Murphy's little brother Hugh who had come along said, "Here's one with a 24 inch blade, but it's a mulching mower. It says its blade spins 20 percent faster than regular mowers."

"I think we should buy the one that lets me finish the yard in only eight-tenths of the time it takes me now," Murphy said.

"Which one is that?" asked Hugh.

"Having the blade turn faster doesn't mean you mow faster," Murphy said. "The issue is how long it takes to run the mower over the entire lawn. As I push our lawnmower, it cuts 24 inches of grass across, where the mower with the 30-inch blade would cut 30 inches. So I'm cutting 24-30ths of what the bigger mower would cut—as a fraction, that's 24 over 30. Reduce the fraction by dividing both the numerator and denominator—the number on top and the number on bottom—by three, and that leaves a fraction of eight over ten—eight-tenths."

20 Don't Fence Me In

Cindy and Luke's family had just adopted a beagle named Trevor from an animal shelter. One thing they soon learned is that beagles go wherever their sense of smell leads them. Trevor was not so much a dog with a nose attached as a nose with a dog attached.

The family quickly realized that they needed to do something about the fence around their back yard. Its slats were so far apart that Trevor could slip between them and get out. Their parents decided that rather than replace the entire fence, they would just attach chicken wire along the bottom.

Before they left for the hardware store with their mother, Cindy and Luke measured the yard. It was a rectangle. Eight feet of fence ran out on each side from the back of the house, which was 33 feet wide. The fence along the sides of the yard was 60 feet long.

At the store, they saw that chicken wire came in rolls of 40 feet.

"We'll need six of these," Cindy said.

"No, that's too much," Luke said.

"You wanna bet?" Cindy asked.

“We need to enclose the perimeter of our yard,” Cindy said, “To find a perimeter, you add the length of the sides. Two of the sides are 60 feet each–120 feet in all. The other two sides also are equal to each other, since the yard is a rectangle. The length of that side is the width of the house, 33 feet, plus 8 feet on each side. That’s 49 feet—33 plus 8 plus 8—so those two sides are 98 feet in total—49 plus 49. Adding the 98 feet of those two sides to the 120 feet of the other two sides means the perimeter of the yard is 218 feet. Five rolls at 40 feet each would be only 200 feet, so we would need a sixth roll for the extra 18 feet.”

“There’s only one thing wrong with that,” Luke said. “We only need to run the chicken wire along the fence, not along the house too. So you have to subtract the width of the house from that; 218 minus 33 is 185. So five rolls at 40 feet each, making 200 feet, would be enough.”

They bought the chicken wire, attached it to the fence and spent the rest of the afternoon playing with Trevor. “A back yard is never complete without a dog,” their mother said.

21 Slow Boat

"You guys will love this place," Jesse's grandfather said. "I go there all the time."

Jesse and his friend Thomas were visiting Jesse's grandfather, who loved boating and had just bought a new boat. From Jesse's grandfather's house along a river, they would be going downstream to a park that had a fishing pier and a restaurant.

"How far is it?" Thomas asked as they got on the boat.

"Twenty nautical miles," Jesse's grandfather said. "A nautical mile is about one-sixth longer than a land mile, or almost two kilometers."

The boat's speedometer showed the engine was running at 20 knots—which Jesse's grandfather said meant 20 nautical miles per hour—on the trip there, which lasted 48 minutes. They had a fun day, although they didn't catch any fish. The speedometer showed the same speed on the trip back, which took one hour and twenty minutes.

Thomas and Jesse climbed out onto the dock while Jesse's grandfather tied up the boat.

"Well, we learned something today," Thomas said to Jesse.

"You mean that we're no good at fishing?" Jesse said.

"Not that. If the distance was 20 nautical miles and the boat was moving at 20 knots, the ride should have taken one hour but it didn't—in either direction. Either the speedometer is wrong about the speed, or your grandfather is wrong about the distance," Thomas said. "But which?"

"They're both right," Jesse said. "What we learned is the speed of the current in the river. On our way there, we were going downstream. Set up an equation: 20 nautical miles in 48 minutes equals an unknown number of nautical miles in 60 minutes—20/48 equals x/60. Reduce 20/48 by dividing both the numerator and denominator by 4, and you have 5/12 equals x/60. To solve, multiply the numerator of each fraction by the denominator of the other, so you have 12x equals 300. To find x, divide both sides by 12. That means x, the speed we were going, is 25 knots. That's five faster than the 20 on the speedometer, meaning the current was pushing us along at five knots."

"I see," Thomas said. "And on the way back, we needed one hour and twenty minutes—that's 80 minutes—to travel the same 20 nautical miles with the engine set at the same speed of 20 knots. As a formula, that's 20 nautical miles in 80 minutes equals an unknown number of nautical miles in 60 minutes—20/80 equals x/60. Reduce 20/80 to one fourth by dividing both the numerator and denominator by 20. To solve, multiply the numerator of each fraction by the denominator of the other, so you have 4x equals 60. To solve for x, divide both sides by four, and x, our speed going upstream, was 15 knots. That's 5 slower than the 20 on the speedometer, meaning the current was working against us at 5 knots."

22 Stepping Up to the Challenge

"Andrew, can you help us?" his little brother Ryan asked.

Ryan and his friend Carlos had been playing with Ryan's toy airplane, and it had landed on the roof of the shed in their back yard.

The roof was 10 feet off the ground and was flat with a ledge around it, so someone had to climb up and reach for the airplane. They couldn't just throw a rope onto the roof to get it down. There were bushes and flowers all around the shed in a flowerbed coming out 5 feet from around the shed. Their mother worked hard to make the flower garden look good.

"I'll need a ladder," Andrew said.

Ryan went into the shed and pulled out a stepladder that was 6 feet tall when it opened.

"There are two more ladders in here," Carlos said from inside the shed. "But they're just straight ladders, not the kind that open up or extend. One's label says it's 10 feet long and the other 12 feet."

"The top end of one of those would have to reach over the top of the shed so it won't slip down the wall," Andrew said.

"You can reach onto the roof climbing up the 6-foot ladder, can't you? I mean, you're more than 5 feet tall, plus the length of your arm," Ryan said to Andrew. "Should we bother getting out another ladder?"

"Yes, we have to get another ladder," Andrew said. "The only way I could reach the roof by using the six-foot ladder would be to put the base in Mom's flower bed."

"That's not a good idea," Ryan agreed.

"So, that means we have to use one of the straight ladders and lean it against the top of the shed. Have you learned about the Pythagorean Theorem yet?"

"No, what is it?" Ryan asked.

"It's how to find the third side of a right triangle when you know the other two sides," Andrew said. "The square of the length of the hypotenuse—that's the long side—is equal to the sum of the squares of the other two sides. Here," Andrew continued, "we have a right triangle whose three sides are the height of the shed, the distance from the shed to the bottom of the ladder—those two make up the right angle—and the length of the ladder from that point to the top of the shed, which is the hypotenuse.

"The bottom of the ladder will have to be 5 feet from the wall of the shed, so it's not in the flowerbed. We'll have a right triangle whose base is 5 feet and whose height is 10 feet. The square of 5 is 25, and the square of 10 is 100. So the sum of those is 125. That means the square of the length of the ladder has to be at least 125 to reach over the edge of the roof. Obviously, the 10-foot ladder is not long enough to act as the hypotenuse because the square of 10 is 100, as I just said. The square of 12 is 144, so the 12-foot ladder will be long enough. You two bring it out and then hold it steady while I climb up."

Getting a Lift

Jada and Michelle's school was closed for a winter teacher training day, so their parents decided to take a day off from work to take the family skiing. They were glad to see when they got there that there were no lines at the chair lifts.

The two girls were good skiers, so they headed to the part of the mountain with the black diamond trails, the hardest ones. Three lifts started next to each other and ran up the mountain, to a spot on the top leading to many different trails.

"Let's try to get in as many runs as we can," Jada said.

They looked at a sign to decide which lift to use. The Sheer Drop lift had four seats per chair and its capacity was 1,200 skiers an hour. The Hang onto Your Hat lift was a two-seat lift with a capacity of 800 skiers an hour. The White Cliffs lift was a three-seat lift that could move 900 skiers an hour. The sign said each had the same number of chairs.

"Where do you think should we go?" Michelle asked.

“Sheer Drop. It moves the fastest—it carries 1,200 skiers an hour versus 900 and 800 for the other two,” Jada said.

“It carries the most skiers but that doesn’t mean it moves the fastest,” Michelle said. “Since there are no lines at the lifts, and all three lifts have the same number of chairs and start and end next to each other, the question is how frequently a lift drops off groups of skiers–in other words, how fast a chair gets from the bottom of the mountain to the top.

“Now, the Sheer Drop lift has four seats per chair and it has a capacity of 1,200 skiers an hour, meaning it makes 300 drops an hour—1,200 divided by four,” Michelle said. “And the White Cliffs lift is a three-seat lift that can drop off 900 skiers per hour, meaning it also makes 300 drops per hour—900 divided by three. The Hang onto Your Hat lift can drop off 800 skiers an hour and has two seats per chair, meaning it makes 400 drops an hour—800 divided by two. So the Hang onto Your Hat lift will get us to the top the fastest.”

24 Shoe on the Other Foot

Tyler and Jordan's family was staying at a dude ranch, where part of the experience for the guests was taking part in the daily chores. Of course, just like at home, neither boy wanted to do more work than the other.

That morning they were restocking the barn. Bags of oats and a crate of horseshoes had to be moved up to the loft. A rope that had baskets at both ends hung from a pulley attached above a loading window high up in the barn wall.

"One of you loads a basket and then raises it," Jerry, the ranch hand, explained. "I'll stand in the loft and pull in the basket. Meanwhile the other one of you fills the other basket. I'll push the first basket out after I empty it, then you send up the other basket. Keep alternating like that. Tyler, you load the oats and Jordan, you load the horseshoes. Now give me a couple of minutes to get up to the loft," he said and walked away.

Tyler looked at the bags of oats and saw they weighed 20 pounds each. There was no weight marking on the horseshoes, which were lying loose in the crate.

"That's not fair. My job is a lot harder. Each bag is way heavier than a horseshoe," Tyler said.

Jordan said, "Yeah, but I'll have to load all these horseshoes eventually, and we don't know how much they weigh. My job could turn out to be harder." He thought for a moment. "Wait, I know a way to make it equal."

"How?" Tyler asked.

“We’ll use the pulley as a balance scale,” Jordan said. “We’ll tie a knot in the rope to shorten it so both baskets are hanging off the ground. Then we’ll put a bag of oats in one of them, and add horseshoes to the other until the baskets are in balance. That will tell us how many horseshoes equal 20 pounds. Then we untie the knot and start sending up filled baskets. If there are either horseshoes or oats left after we’ve loaded all of the other, we’ll take turns.”

The Hole Truth

The students were helping to create a nature area alongside their school, where plants native to the state would be grown. The younger classes were planting grass and small shrubs, but Mrs. Santorelli's 8th grade class had been assigned to the harder jobs.

Most of the class was digging holes to plant trees, while several students were assigned to dig out the area for a plaque that was being donated by the parent-teacher association. The plaque was going to be a six-foot wide circle made of bronze. On it, there would be a relief map of their state showing the mountains, rivers, and other natural features.

The hole had to be prepared now, before the plaque arrived, so that workers could pour the cement and put it in place as soon as it got there. Karen and Ethan were given the job of digging a hole one foot deep and one-third larger than the plaque.

"We need to make the hole eight feet wide, since the plaque will be 6 feet across and one-third of six is two," Karen said.

"No, we don't," said Ethan, who was going to take the first turn with the shovel.

"I think you're just trying to get out of doing your share of the work," Karen said.

"Am I?" he asked.

"Do some estimating," Ethan said. "The plaque is going to be a circle 6 feet across, which means it has a radius of 3 feet. The area of a circle is pi times the square of the radius, which in this case is 9. Use 3.14 as the value of pi. Multiplying that by 9 means the plaque is about 28 square feet. We need a hole one-third larger than that, so to be safe, let's say 10 square feet more, or around 38 square feet total.

"Now, if we made the hole 8 feet across like you suggest, the radius would be 4 feet, so the square of that would be 16. Multiply that by 3.14 and you have a hole of around 50 square feet. That is a lot bigger than we need.

"Let's say we made the hole just 7 feet across instead. The radius would be 3.5 feet, and the square of that is about 12. Multiply that by 3.14 and we're right about at the 38 square feet we need," he said.

"Let me see for sure," Karen said, doing the multiplication on a calculator. "The plaque will be 28.26 square feet—9 times 3.14. A third of that is 9.33, so the hole needs to be 37.59 square feet. If we dig a hole 7 feet across, the radius would be 3.5, and the square of 3.5 is 12.25. Multiply by 3.14 and the area is 38.465 square feet. So you're right, a 7-foot wide hole will be big enough."

"Doing a little math is much less work than digging a hole a lot bigger than it has to be," Ethan said as he took the first shovelful.

26 In the Deep End

Sydney and Julia were going to start their first summer jobs as lifeguards at the neighborhood pool in a few weeks. For now, their job was to help get the pool ready. All the employees had spent several days cleaning up the pool house and deck, and now they were getting the pool ready to be filled with water. Over the winter, a large pile of leaves had accumulated at the bottom of the pool.

Unfortunately, all they had to take the leaves away were some packing boxes. A couple of the older lifeguards, Destiny and Victoria, had been working together with a two-foot cube box, when the head lifeguard found a three-foot cube box and told Sydney and Julia to help out and use that.

"Fill your box all the way to the top like we've been doing," Destiny advised them. "That way you won't have to climb up out of the pool as often to dump the leaves."

The two teams of girls worked for about an hour, counting how often they filled up their boxes. Finally they finished.

"That was 50 times for us," Victoria said.

"Julia and I filled ours 20 times," Sydney said.

"Well, then it's obvious who carried out more leaves, isn't it?" Destiny said.

"Destiny and I carried out more leaves," Victoria said. "We filled our 2-foot box 50 times and you filled your 3-foot box only 20 times. Our box has a base of 4 square feet—2 times 2—and we filled it 50 times, so that's 200. Your box has a base of 9 square feet—3 times 3—and 9 times 20 is 180."

Julia said, "The issue here isn't the area of the box's base, it's the box's volume. That means we need to know the cubic feet of the box, not the square feet of its base. A 2-foot cube holds 8 cubic feet—the width of 2 times the length of 2 times the height of 2. A 3-foot cube holds 27 cubic feet—3 times 3 times 3. Since both pairs of us filled our boxes to the top each time, you two carried out 8 cubic feet 50 times—400 cubic feet. Sydney and I carried out 27 cubic feet 20 times—540 cubic feet. So we carried more."

27 A Ton of Trouble

Spencer was big for his age, he was both taller and heavier than his friends. This was no surprise, because his father was a big man. Everyone thought that Spencer had a future as a football player.

Spencer and Marco were talking about that as they rode in Spencer's family car on their way to a week at camp. They had loaded their sleeping bags, knapsacks, and other supplies in the back. They had been driving a long time when they saw a sign showing that the camp was still 25 miles ahead.

"I used to go to this camp every summer. I know a shortcut over the mountain," Spencer's father said. He turned off the main road onto a gravel road.

After a few miles they saw a sign that said: "Warning: Wooden Bridge Ahead. Weight Limit 1 Ton."

Spencer's father stopped the car, pulled a small book out of the glove compartment and studied it for a few moments.

"The owner's manual says that the car weighs 1,935 pounds," he said. "So it should be okay."

"I don't know, Dad," Spencer said. "We've got an awful lot of stuff in the car."

"I know what we should do," Marco said.

"What?" Spencer asked.

"Oh, I know what you're going to say," Spencer said. "We should lighten the car by taking out our supplies and carrying them across. One ton is 2,000 pounds, and if the car weighs 1,935 pounds, that's only 65 pounds to spare, and our stuff probably weighs more than that. We have to unload the car."

"No, we have to go back and take the long way to camp," Marco said. "Even if we did what you suggested, your father still has to drive the car. There's no way he weighs less than 65 pounds."

28 Go Take a Hike

Carla and Amanda's family was vacationing at a national park one summer and decided to take a hike down from the top of a gorge to see the river below. A sign said:

Three trails lead from here to different points along the river. The trails do not join each other, and each takes approximately two hours to walk.

Riverside Trail: Steepest. Plan on taking 30 minutes down, 1 1/2 hours back.

Scenic Overlook Trail: Medium steepness. Plan on taking 40 minutes down, 1 hour 20 minutes back.

Forest Path: Most level. Plan on taking 1 hour down, 1 hour back.

Caution: No water available on the trails. Do not drink water from the river or any streams along the way. Please carry water and use it wisely.

They saw another family that had just finished a hike.

"How was it?" Carla asked.

"It was great," the other family's mother said. "But take that warning about water seriously. We'd used one-third of our water when we got to the bottom, and that was just right."

"What trail did you take?" Amanda asked.

"They took the Scenic Overlook Trail," Carla said. "The Forest Path takes the same amount of time to walk back as to walk down—one hour each way. When you reach the bottom of that trail, you've walked one half of the total—one hour out of two hours. The Riverside Trail takes three times as long to walk back as to walk down—90 minutes back versus 30 minutes down. When you reach the bottom of that trail, you've walked one quarter of the total—30 minutes out of 120. The Scenic Overlook Trail takes twice as long to walk back as to walk down—80 minutes back versus 40 minutes down. So when you reach the bottom of that trail, you've walked for one third of the total—40 minutes out of 120."

29 Chute in the Works

On Saturday morning, Caleb rode up the bike path to his friend Patrick's house. That morning Patrick was in his yard painting a model rocket. As much as Caleb loved bicycle riding—he had a bike with a speedometer, lights, water bottle holder and other accessories—Patrick loved model rockets.

"Cool," Caleb said, admiring his friend's new rocket.

"Maybe too cool to use," Patrick said.

Patrick and his father belonged to a club that launched model rockets. Sometimes, though, rockets crashed and broke apart because their parachutes didn't open. The parachute for Patrick's new rocket was already attached to the nose cone.

"I'm worried about this parachute," Patrick said. "The instructions say it should open when the rocket hits 30 miles an hour on the descent. I've tried to test it, but I guess I can't throw the nose cone that fast."

"I'll take it on a ride down the bike path," Caleb suggested. "Once I get going that fast, we'll know if it will open or not."

Caleb tried several times but could never get the parachute to open.

"Sorry, I can't get this bicycle going more than about 20 miles an hour," Caleb said when he returned.

"I have an idea," Patrick said. "And I wouldn't suggest this if you weren't a good enough biker to handle it."

"What do you have in mind?" Caleb asked.

"Once you get going on your bike, throw the nose cone forward," Patrick said. "The speed of the throw will be added to the speed of the bicycle. So if you're riding at 20 miles an hour and you throw it at even just 10 miles an hour, it will be moving forward at 30 miles an hour, and you'll see if it opens."

Caleb did just that. It was a little tricky since he had to steer with only one hand and throw with the other, but it worked. The parachute opened.

"Now it's time for lift-off!" Patrick said.

30 How Much Wood?

Each day at camp started with a recording of a trumpet blaring "Reveille" over the loudspeakers. After breakfast, campers would clean up their cabins, and then, on most days, the morning activities started.

Today, however, was chores morning, and activities wouldn't start until all the chores were done. An announcement after clean-up time told each work group what their assignment was. The chores rotated, and it was Jake and Manuel's turn to carry the firewood for their cluster of five cabins. They were too far up into the woods to be reached by a vehicle, so they would have to carry the wood up the hill from the large stack near the main campfire circle.

"That loudspeaker is so scratchy, I can't understand half of what they're saying," Jake said as the two of them walked down the path. "What did they say about the firewood?"

"They said the log carriers for each cluster have to bring up logs and divide them evenly among the cabins in their cluster," Manuel said.

"Okay, but how many logs total?" Jake asked. "It was either 13 or 30—I couldn't make it out."

"Do you have to?" Manuel asked.

“If we have to divide the number of logs evenly among these cabins, it means we need to bring 30 logs,” Manuel said. “Thirteen is a prime number, like three or seven. That means that it’s not divisible without a remainder by anything other than itself and one.

“So, since there are five cabins in our cluster, we need to bring 6 logs to each one—30 divided by five,” he explained.

Math at Play

31 Jumping Through Hoops

Ms. O'Cork, the girls' P.E. teacher, tried to mix up the activities to give her class different kinds of exercise.

Today, she had brought out a bunch of hula hoops for warm-ups, which the girls enjoyed.

They were out in the back field, which had no distance markings because it was used for all kinds of sports.

After warm-ups, Ms. O'Cork gathered everyone on the edge of the field, where she dropped a large bag of soccer balls and some short tape measures, the kind used to measure people for clothes.

"The school record for punting a soccer ball is 45 yards," she announced. "Anyone who can break the record in the next two minutes and can prove it doesn't have to run laps later."

"But it will take that long just to measure 45 yards with these little tape measures," someone said.

"Okay, anyone who can figure out how to accurately measure the distance in that time doesn't have to run laps either," the teacher said.

Jasmine turned to Audrey, who was the goalie on their soccer team and a good punter. "I know a way we'll both get out of running laps," Jasmine said.

"What do you have in mind?" asked Audrey.

Jasmine used the tape measure to measure the circumference of a hula hoop, starting at the joint where the two ends joined. It was 108 inches around.

"Dividing 108 inches by 12 inches, the equivalent of one foot, means the hula hoop is nine feet around, or three yards," Jasmine said. "I'll roll it along the ground. Every time the joint comes around is three yards farther. Forty-five yards divided by three yards per roll is 15. So after 15 rolls of the hula hoop, I'll be 45 yards away. Now warm up that kicking leg!"

32 Ace of Clubs

Natalie and her father had been taking golf lessons. They were hitting the ball pretty well, so they thought it was time to go out and play their first real round of golf.

On the first hole, they hit their drives down the fairway.

"This marker says we're 150 yards out from the green, Daddy," Natalie said when they reached his ball.

"Okay, the instructor said 150 yards is how far I hit with a six-iron," her father said, pulling out that club. He took a practice swing that was interrupted when his hat flew off back toward the tee, making Natalie laugh.

He hit the shot the way he usually did, but it landed 30 yards short of the green. "I could have sworn he told me I hit six-irons 150 yards," he said.

The next hole ran parallel to that one, but going the other way. After their drives, Natalie's father was once again about 150 yards from the green. "Let's see, the instructor said there's about a 15-yard difference in how far different clubs send the ball, and the lower the number of the club the farther the ball goes. So if I hit the six-iron 120 yards like I did on the last hole, I'll need to use the longer club that will hit it 30 more yards. That means a four-iron," he said.

"I wouldn't do that if I were you, Daddy," Natalie said.

"Why not?" he asked.

"On the first hole you hit a shot that normally would travel about 150 yards," she said. "That shot was into the wind. You hit a good shot, but it still only went 120 yards. So, the wind reduced the distance of your shot by 30 yards, or a fifth.

"On this hole, we're going the opposite direction, meaning the wind is behind us. So the wind will add about one-fifth to the distance of your shot. So hit the club that normally makes the ball go about 120 yards, and let the wind push it. Since you usually hit the six-iron 150 yards, and each higher numbered club sends the ball 15 yards less, you should use an eight-iron."

33 A Slice of Life

After practice one day, some of the divers decided to hang around the pool. None of them had eaten for several hours, and everyone was getting hungry.

"Who wants pizza?" their coach Justin asked. There was a delivery place just down the street.

Everybody's hand shot up.

"And what do you want on it?"

Suggestions came flying in—pepperoni, mushrooms, olives, anchovies. Others just wanted cheese.

"Okay, I count three people who want it plain, and four who want a topping. I'll just have to be careful with how I order it," Justin said.

Less than half an hour later, two large pizzas were on the picnic table. Although they were the same size, they looked different. The plain one had been cut into six slices and the one with toppings into 12 slices, with each of the four toppings on three of the slices.

Anna's little brother Matt was one of the three kids who wanted the plain pizza. He and the other two quickly ate their two slices each and watched as Anna ate her third slice with anchovies.

"No fair!" he said. "You got three slices and I only got two."

"True, but you ate more pizza," Anna said.

"No fair!" Matt repeated. "You ate three, and I only ate two. How can you say I ate more?"

“Both pizzas were the same size,” she reminded him. “The plain one was cut into six pieces and you ate two slices, which is two sixths, or one third. The one with toppings was cut into 12 slices, and I ate three slices, which is three twelfths, or one fourth. Your one third is bigger than the one fourth I ate.”

“I still think you’re cheating,” he said.

“Look at it this way,” she answered. “Let’s say your pizza had been cut into 12 slices like ours was. That means each of the six slices would have been cut into two halves. So you ate an amount equal to four of the slices on my pizza. You ate four twelfths while I ate only three twelfths.”

34 A Perfect 10

"What's the qualifying score for States again?" Alison asked.

"A 33.5," Emily told her.

The two gymnasts were at a sectional meet, which served as a qualifier for the state championships. Each was competing in all four events—vault, bars, beam and floor—with a possible top score of 10 in each event. Both of them were trying to get at least the qualifying score.

They had been planning to be roommates at States if each qualified. After two rotations, bars and beam, Emily had a total score of 17.25, but Alison had just a 16.5.

"I'll never make it," Alison said glumly. "I guess you'll have the room all to yourself."

"You can do it," Emily said. "We still have floor and vault to go, and you do well on those. What's your average on them?

"An 8.75 on floor and 8.25 on vault," Alison said. "What's your average on them?"

"An 8 on each," Emily said.

"Well, one of us will only have to be average, but the other one will have to do better than average if we're going to share that room," Alison said.

"Which one, you or me?" asked Emily.

At the end of the awards ceremony, the girls walked off the podium. Alison had finished with a 33.5 all-around score, exactly what a gymnast needed to advance to the state championships, while Emily had also qualified, with a 34.25.

Alison said, "After two events, I had a 16.5, which meant I needed a total of 17 on the last two events to make it—33.5 minus 16.5 is 17. My average scores on those events were 8.75 and 8.25, which added together make 17. So I knew I'd get a 33.5 if I got my average scores."

Alison continued, "After two rotations you had 17.25 points, which meant you needed 16.25 more points—33.5 minus 17.25 is 16.25. But your average on the last two events was 8 each, meaning an average total of 16—8 plus 8. You were the one who needed to do better than your average on the last two events to make it to States. I didn't want to say that and make you nervous. I'm glad you made it. We're going to have a great time there!"

35 Cutting Corners

"Who wants to take drinks to the older boys?" Brandon's father asked.

Voices called out: "Me! No, me! Me! I want to!"

Many of the players on Brandon's soccer team had older brothers on the team that had just finished the first half of their game. Brandon's team was going to play a game on the same field afterward, and his father, the coach, suggested that the entire team come early to watch the older kids play.

It was a hot day, so Brandon's team was behind a corner of the field in the shade. One of the parents had brought a case of sports drinks for both teams to share. The older team was going off to the opposite corner of the field, where there was also shade.

"We don't need everybody to go," Brandon's father said. "Ali, Jacob, Christian, Luis and Brandon, how about you take an armload of bottles each?"

I might as well be polite and go around the outside of the field, Brandon said to himself. But then he saw the others cutting across the middle.

"Race you!" Jacob called and they all started running as fast as they could. Brandon continued around the outside of the field and got to the older team last.

"I know I'm faster than them," Brandon said to his older brother Victor as they handed out the drinks. "How did they beat me?"

Victor said, “You were running faster than them—I could see that—but they had less distance to cover. A straight line is always the shortest distance between two points.”

36 Net Result

"Man, we'll never even get a shot off," Zachary grumbled. He and his teammates on their middle school basketball team were reading a newspaper story about the upcoming city championship game. Their team, the Dragons, would be playing the Cheetahs, a team they hadn't played in the regular season.

"It says here that their starting team averages 5 feet, nine inches, and they have one player who is 6 feet, 1 inch, another who is 6 feet, 2 inches, and the other three are the same height as each other," Logan said.

The article went on to say that the Cheetahs would have a big size advantage over the Dragons, whose tallest player, William, was 6 feet tall. The other four starting Dragons, Zachary, Logan, David and Gabriel, all were 5-feet, 7 inches, or 5 feet, 8 inches—each of them shorter than the Cheetah average.

"Actually, I like our chances," Gabriel said.

"How can you say that?" William asked.

Gabriel said, "Their starting team averages 5 feet, 9 inches. That's 69 inches per player—5 feet times 12 inches to the foot is 60, plus 9 is 69. So the total height of their starting five players would be 345 inches—5 times 69 is 345.

"Now, a player who is six feet, one inch accounts for 73 of those inches all by himself—six feet times 12 inches to the foot is 72, plus one is 73. The player who is one inch taller is one inch more, or 74. So those two guys account for 147 inches of their team's total height—73 plus 74. That means their other three players add up to only 198 inches—345 minus 147. Since the article says they're all the same height, each of them must be 5 feet 6—198 inches divided by three is 66 inches, which is the same as 5 feet, 6 inches. All of our players are at least 5 feet 7. So three of us will have a height advantage compared to only two of them."

37 Capture the Difference

Capture the flag was a favorite game in P.E. class, and the place to play it was on the blacktop, a large paved area for outdoor games.

There was a line in the middle, and each team had a flag to protect while trying to bring the other team's flag back across the line. Any player on the other team's side of the line could be captured by being touched. Each team could have two players guarding their own flag, but they had to stay a certain distance away from it unless a player from the other team had grabbed it.

It had rained for several days, and all the chalk lines had to be redrawn. The P.E. teacher, Mrs. J, handed Veronica a piece of chalk and a 20-foot long piece of string.

"Use these to draw the area that your team's guards aren't allowed to go in," Mrs. J said. "The area has to be as big across as the string is long."

"Okay," Veronica said. "A circle or a square?"

"The kids on the other team made a square," the teacher said. "But as I told them, you can do it however you want. Take a minute and think it through."

"I don't need to. I know what I should do," Veronica said.

"How did you decide?" Mrs. J asked

"It's just a question of area," Veronica said. "A square that's 20 feet on each side would have an area of 400 square feet—the length of 20 times the width of 20. The area of a circle is the square of the radius times pi. The radius is half of the diameter, so if the diameter is 20, the radius is 10, and the square of that is 100. We'll use 3.14 as the value of pi. When you multiply a number by 100, you just move the decimal point two places to the right. So that's 314 square feet, which is less space for our guards to defend."

38 Way to Go

Kyle and his friend Antonio had been watching the summer Olympics the day before. Their favorite events were track and field. Kyle was the fastest sprinter in school, while Antonio liked running longer distances.

On this day, they were at the high school football stadium, practicing for the upcoming fall track season. Antonio warmed up by running a few laps while Kyle stretched and ran some short sprints.

Kyle gave Antonio his stopwatch and asked Antonio to time him as he sprinted from one goal line to the other. Antonio timed him in 15 seconds flat.

"Olympic gold medal, here I come!" Kyle yelled.

"How do you figure that?" Antonio asked.

"Remember when we were watching the sprinters yesterday, the winner in the 100-meter dash was just under 10 seconds? All I have to do is run about one-third faster. Five seconds is one-third of 15 seconds, and if I cut off those 5 seconds, I'll be right around 10 seconds. There's an Olympics every four years. I bet that eight years from now I'll be one-third faster."

"I hope you do win an Olympic gold medal someday," Antonio said. "But you won't need to run a third faster."

"Great!" Kyle said.

"Actually, not so great," Antonio said.

"What do you mean?" Kyle asked.

"Unfortunately, you'll need to run more than one-third faster," Antonio said. "You ran a shorter distance in 15 seconds than they did in 10 seconds. The markings on a football field are in yards, and the distance from one goal line to the other is 100 yards. As you said, the Olympic events measure distance in meters, which are longer than yards. One meter is a little more than 1.09 yards, a little more than nine percent longer. So let's time you again going 109 yards—that's about equal to 100 meters—and you'll have a better idea of how much faster you'll need to go to get that gold medal."

39 Hit Parade

"Four for four! All three of us!" Richard said.

He, Miguel and Colin were celebrating after winning a baseball game thanks to their good hitting. All three of them had gotten hits in every one of their at-bats. Under the league's rules, starting players could have no more than four at-bats, and since this was their sixth game of the season, each of them had had twenty at-bats going into the game.

Before the game, Miguel had a batting average of .400, Richard of .250 and Colin of .200.

Richard said, "Miguel, this game really helped your batting average."

Miguel said, "Are you sure it didn't help you and Colin more?"

"How could it?" Colin asked. "You have the highest average of any of us!"

"A batting average is the number of hits divided by the number of at-bats," Miguel said. "First, you need to know how many hits we had before today's game, which you can figure out by knowing our batting averages and the fact that each of us had 20 at-bats before today. My .400 batting average before today means I got a hit in 4 out of every 10 at-bats—that's 8 hits in the 20 at-bats before today. A .250 average means a hit in 2.5 out of every 10 at-bats, which means Richard had 5 hits before today. A .200 average means 2 hits out of every 10 at-bats, which means Colin had 4 hits before today."

"I get it," Richard said. "So after today's game, we each had 4 more at-bats, making 24, and 4 more hits each. That means Miguel now has 12 hits out of 24 at-bats, or 1 hit out of every 2 at-bats. To put it in decimal terms as batting average, that's 1 divided by 2, or 500. He raised his batting average from .400 to .500 today, 100 points."

Richard continued, "I now have 9 hits in 24 at-bats, or 3 in every 8, and 3 divided by 8 is .375. I raised my batting average today by 125 points, from .250 to .375. Colin now has 8 hits in 24 at-bats, or 1 in every 3, and 1 divided by 3 is .333. His average is now .333 compared to .200 before today. So he actually raised his average the most, 133 points."

40 Miniature Math

"I can't wait to see our new miniature golf hole actually built!" Gianna said. She and her friend, Jason, were one of two teams that had just won a "Design a Miniature Golf Hole" contest. Each team won a cash prize and was having their hole design built at one of the two new miniature golf courses at a theme park.

Each team's hole would be the last one on each course, so the cups would not have bottoms. The ball would go down a pipe and straight to the cashier.

Edward and Sophie made up the other winning team. All of them were at the prize ceremony, which featured the drawings of their holes.

Gianna and Jason's design was a straight path to the hole, but there were rocks and bricks blocking the direct path of the golf ball. Edward and Sophie had created a jagged design without many other obstacles to the hole. Both teams had added a ledge where the ball would drop down to a lower level. The cup would be on the lower level.

Getting the ball to the ledge would be equally challenging on both holes. The difference was that Gianna and Jason's hole would have a 4-inch wide cup, and Edward and Sophie's would have an 8-inch wide cup. There would be prizes for players whose balls dropped in the cup in just one shot.

At the ceremony, one of the judges made an announcement about the prizes. "Since the Gianna-Jason team's cup is only half as wide, it will be twice as hard to make a hole in one. So we will award one free game to anyone who gets a hole-in-one on the Edward-Sophie hole and two free games to anyone who does it on the Gianna-Jason hole."

"That doesn't seem fair," Sophie said to Edward.

"Why not?" he asked.

"If a ball were rolling along the ground, yes it would be twice as hard to get it into a 4-inch-wide cup than into an 8-inch-wide cup," Sophie said. "But the ball will be dropping off a ledge. That means the key issue is the area of the cup, not its diameter. And since the cup won't have a bottom, we don't have to take into consideration anything unusual like the ball landing in the cup and bouncing out."

"The area of a circle is pi times the square of the radius," Gianna said, getting out a notebook to do a calculation. "Our cup with a 4-inch diameter has a 2-inch radius, and 2 squared is 4. Multiply that by 3.14 as the value of pi and the area of our hole will be 12.56 square inches. And their cup with an 8-inch diameter has a 4-inch radius, and 4 squared is 16. Multiply that by 3.14 and the area of their hole will be 50.24 square inches. That's an area 4 times bigger. So if making a hole in one in the larger cup is worth 1 free game, making one in the smaller cup should be worth 4."

41 Calling Long Distance

"I don't know, honey," Mia's mother said when Mia brought home a flyer from school about a mother-daughter charity race. "I haven't jogged since college."

"They don't even call it jogging anymore, Mom. It's running," Mia said. "But you can do it. It's only three miles. I'll run with you." Mia was a good runner and ran on the middle school track team.

Mia's mother decided to try. She got a pedometer, a device that measures how far a runner has gone by counting strides, after the runner enters the length of a typical stride. She figured out the length of her stride by counting her steps while running at the high school track.

For training, she thought that running uphill would help her get in better shape, so Mia's mother worked out a three-mile route from a spot downhill from where they lived. Every other day, Mia's father dropped Mia and her mother off there and they ran home, her mother pushing herself uphill with short, choppy strides while Mia ran smoothly alongside, encouraging her.

On the day of the race, Mia had no problem covering the distance. Her mother managed to finish, but it was a struggle.

"That course seemed really long," Mia's mother panted at the finish line. "Maybe they measured it wrong?"

"You measured our course wrong," Mia said.

"No, I'm sure I set the pedometer correctly," her mother said.

"Yes, but you set it according to the length of your stride running on the level," Mia said. "You ended up putting faulty data into the calculation. When you ran uphill, you used shorter steps than when you ran on the level. That means you were covering a shorter distance per step. So, in our training, we actually were running less than three miles."

"Why didn't you tell me then?"

"I didn't want you to get discouraged. I thought once you got in the race you could go the extra distance, and you did," Mia said.

"If you want," she added, "we could measure out a distance of, say, 100 yards up the hill and count how many strides you take. That would give us the correct length of each stride you take going uphill. Multiplying that by the number of the strides you took when we ran our training route would give us the true distance of that route."

"Maybe some other time," her mother gasped.

Luck of the Draw

To celebrate the end of the grading quarter, Jorge's class had a math game day.

Students divided into pairs. For half of the period, a pair would host a game and, for the other half, they could play other teams' games. Each team got 10 game tokens to play with and a roll of prize ticketss to give out to winners. Whoever won the most tickets would get extra credit on their next test.

Jorge and his friend Kennedy were in a pair. When it was their time to play, they walked around checking out the games.

At one table, Josiah and Brianna had a pair of dice and were offering five tickets for anyone throwing a double.

At another, Kenneth and Daniela had a deck of cards and were offering three tickets for naming a suit and pulling out a card of that suit.

At a third table, Preston and Angel had a bucket with eight toy ducks floating in it. Their sign said that one of the ducks had something written on the bottom and that picking it would win six tickets.

"Aren't all these games just a matter of luck?" Jorge asked.

"Not exactly. There's a way we can improve our chances of winning the most tickets," Kennedy said.

"How?"

"We can compare the payouts of each game," Kennedy said. "Let's start with Josiah and Brianna's dice game. Each die has six sides, so with two dice, there are 36 possible combinations—six times six. Of those, six are doubles—two ones, two twos, and so on. That means the chances of throwing a double are one in six—36 total possibilities divided by the six ways to make a double. Josiah and Brianna are offering five tickets if you do it, so they are offering five-sixths of the value of winning their game."

"There are equal numbers of cards of each of the four suits in a deck of cards, so the chances of picking the suit you name is one in four," she continued. "Kenneth and Daniela are offering three tickets if you do it, so they are offering three-fourths of the value of winning their game."

"I see," Jorge said. "So in Preston and Angel's duck game, the chances of picking the duck with writing on the bottom are one in eight, and they're offering six tickets if you do it. So they're paying six-eighths of the value of winning their game."

"Now all we have to do is compare the three fractions—five-sixths, three-fourths and six-eighths—by finding the lowest common denominator, the smallest number that each of the denominators is a multiple of," he said. "That would be 24. Five-sixths multiplied top and bottom by four becomes 20-24ths. Three-fourths multiplied top and bottom by six becomes 18-24ths. And six-eighths multiplied top and bottom by three also becomes 18-24ths. So Josiah and Brianna's dice game that pays 20-24ths offers the most value. Let's play that one."

"Of course," Kennedy added, "just because it's most probable that we will win more tickets playing that game doesn't mean that we actually will. But that game gives us the best chance."

43 Head over Heels

Gabrielle, Angelina and Makayla's competitive cheerleading team was holding a cartwheel-a-thon to raise money for a trip. Each girl would ask people to pledge a certain amount of money for every cartwheel she could do in five minutes, up to a maximum number each girl set for herself. A coach would count, and there would be a pizza party at the end.

"Everybody thinks I'm going to make myself dizzy," Angelina said on the day they brought in their pledge sheets.

"My mom says I'm going to be sick to my stomach," Gabrielle said.

"My dad says he won't even be able to watch," Makayla laughed. "How many do you think you can do?"

"I said I'd do no more than 100," Gabrielle said. "I got four sponsors at 10 cents per cartwheel and one at 25 cents."

"I told mine I'd do no more than 12 a minute," Angelina said. "I got three sponsors at 30 cents each."

"My parents donated $30, and I figured I could average one every 10 seconds," Makayla said. "I got five more sponsors at 20 cents each for that."

"Well, I think we can all do as many as we promised," Gabrielle said. "I wonder which of us will raise the most money if we do."

Angelina said, "Gabrielle, your sponsors pledged 65 cents per cartwheel in total—one at 25 cents plus four at 10 cents each. So if you do 100 cartwheels, that would be 65 cents times 100, or $65.

"My sponsors pledged 90 cents per cartwheel in total—three times 30. If I do 12 cartwheels a minute for five minutes, that's 60 cartwheels—12 times five—so I would raise 90 cents times 60, or $54.

"And Makayla, five of your sponsors pledged $1 per cartwheel in total—five times 20. If you do one cartwheel every 10 seconds, that would be six per minute—60 seconds in a minute divided by 10. Over five minutes, that would be 30 cartwheels. So that's $1 times 30, or $30. And when you add the $30 your parents pledged, that would be $60. So Gabrielle will collect the most—if she really can do 100 cartwheels in five minutes."

"I'm sure I can do them," Gabrielle said. "But I'm not sure I'll be able to eat that pizza afterward."

44 Batter Up

Mr. Andropolous sent the boys' P.E. class to the baseball field, but it didn't seem like they were going to play a game. There were no gloves—only a bag of balls, a couple of bats, several tape measures and a batting tee near home plate. Chalk lines had been drawn in arcs across the outfield, starting just beyond the infield.

"Those lines are for the school fair tomorrow," the teacher said. "There's going to be a homerun-hitting contest, only with batting tees. People will get a ticket for a prize if they hit it past a certain line. The woman running the game wants to give a ticket to half the people who try."

"So you're going to use us to decide which line they have to hit it past?" Michael asked.

"That's right," said Mr. A. "I think this group is about average. Line up, hit one ball each and go to the back of the line. When all the balls have been hit, we'll measure how far each one went, using those lines and the tape measures. We'll add the distances, divide by the number of balls, and that will tell us how far the average ball went. The closest line to that will be the target line for the fair."

"I hate to make a bad joke, but I can think of a more 'fair' way to do this," Michael said.

"Let's hear it," Mr. Andropolous said.

"If you're trying to find the distance that half of the people will hit the ball past, you're looking for the median distance, not the average distance," Michael said. "The median is the number that half of a sample are above and half are below. So after we hit the balls, we need to find the distance that half are beyond and half are short of. That's where the target line should be."

45 Doing Swimmingly

"We got spirit, yes we do, we got spirit how 'bout you?"

The swimming team always divided into three groups by age before practice and did that cheer one group after the other. It helped them get hyped up for outdoor practice early in the mornings when the water was chilly. But on this first day of practice for the season, the paper that the coaches handed out was even more of a shock than the cold water.

It said:

Team Goals:
Have fun. Give your best. Be a good sport, win or lose.

Individual Goals:
Ages 14 and up, 10 percent improvement in personal best time
Ages 11-13, 20 percent improvement
Under age 11, 30 percent improvement

Andrea, Madeline and Leah were sisters who specialized in the 100 meter freestyle swim. Andrea was 14, and her personal best time was one minute, four seconds. Madeline, who was 12, had a personal best time of one minute, twenty seconds. Leah, who was 10, had a personal best of one minute, thirty seconds. They were very competitive about their swimming.

"You know, this could make somebody pretty mad," Andrea said.

"You mean Leah will get mad because she has to improve by the most?" Madeline asked.

"Not quite," Andrea said.

"To see how much they want us to improve, first convert our personal best times into seconds," Andrea said. "My best time of one minute and four seconds is 64 seconds—60 seconds to the minute, plus four. My age group's goal is to improve by 10 percent. To find 10 percent of a number, move the decimal point one place to the left. So 10 percent of 64 is 6.4. If I meet the goal, my time will be 64 seconds minus 6.4, or 57.6 seconds."

"Madeline's personal best of one minute and twenty seconds is 80 seconds. A quick way to find 20 percent of a number is to first find 10 percent and then double the result. So 10 percent of 80 is 8, and double that is 16. If Madeline meets her goal, her time will be 80 seconds minus 16, or 64 seconds."

Andrea continued, "Leah's personal best of one minute and thirty seconds is 90 seconds. Ten percent of that is 9 seconds, so to find 30 percent, triple the result—27. And since 90 minus 27 is 63, if both of you meet your goals, Leah will beat Madeline's time by one second."

Andrea and Leah laughed, but Madeline dove into the pool and started swimming, very hard.

Math Every Day

46 Rows and Columns

"Man, how did those ancient Greeks build those temples? It's hard enough just drawing one," Lucas said as Deandre walked into the auditorium.

Their class was in a unit on ancient civilizations. Once they had gone through the unit, they were going to put on a play about it. Lucas and Deandre were in a group assigned to paint the backdrop, a piece of white cloth 10 feet high by 15 feet wide that was now lying on the stage.

Their teacher, Mr. Gage, had given them a poster, two feet high by three feet wide, of an ancient Greek temple for them to use as a model. Before starting to paint the backdrop, the students had decided to sketch on it with chalk, to give them a guide. They were using yardsticks to help make the lines straight.

"It does look wrong," said Deandre, who had been in the art room collecting paints while the others sketched. "The columns all are the wrong length or width, and the roof looks funny. Why don't all of you take a break and let me try it?"

The other students wiped off the chalk marks, then went back to the classroom for a while. When they returned, Deandre had filled the backdrop with a sketch of the temple that looked just right.

"How did you do that?" Lucas asked.

"It's just a matter of extrapolating," Deandre said.

"Extrapolating?" Lucas repeated.

"Using figures you know to get figures you need to know," Deandre said. "The proportions on the backdrop were the same as those on the poster—a ratio of 2 in height for 3 in width. Since the backdrop is 10 feet high and the poster is 2 feet high, anything on the poster has to be made 5 times longer on the backdrop to keep the proportions the same, because 10 is 5 times larger than 2. So, a column that measured 10 inches tall on the poster, for example, had to be 50 inches tall on the backdrop. Now all we have to do is get going with the paints."

47 Sweet Solution

"Come on, class, to the candy aisle!" Miss Hanson called out.

The class groaned at the thought of being in the candy aisle, knowing that they couldn't have any for themselves. They were on a field trip to a grocery store to buy supplies for a project.

They had collected pine cones, and they were going to cover them with peanut butter, roll them in bird seed, stick on candy, and tie strings on the pine cones to hang in the trees outside the school.

Patti was in charge of the candy. "So tell me again. What do I have to do, Miss Hanson?" she asked.

"Well, Patti, remember how we collected 200 pinecones? We are going to put five peppermints on each pine cone for the squirrels, so start counting while I take the other students to get the peanut butter and bird seed."

"That's 1,000 pieces of candy," Patti's friend Lulu said as they reached the candy. The peppermints, each wrapped in plastic, were in a bin and were sold by weight. There was a scoop to use with a scale.

"Too bad they don't sell it in bags marked with how many are inside," Lulu said. "It will take you forever to count them."

"Actually, it will be a lot easier than I thought," Patti said.

"What do you mean?" Lulu asked.

"This scale will be a big help," Patti said. "I'll count out a sample of the candies and weigh them. I'll count out, for example, 50 candies, and once I know their weight I'll put them in the bag. Then I can just use the scooper. When I've scooped enough candies onto the scale to produce an equal weight, I'll know I have an equal number of candies. Then it's just a matter of repeating until I get to 1,000."

48 Driving Them Crazy

"Are we there yet?" Darius asked from the back seat.

Axel's little brother had asked that so many times that Axel—who was eager to be finished with the trip himself—had taken to timing how long it was since the last time Darius asked. Axel's watch had a stopwatch feature and was accurate to the one-hundredth of a second. They were on their way to a week at the beach, and the trip was a long one.

"How long was it this time?" their father asked from the front seat.

"Six minutes, 38.55 seconds," Axel responded.

"We're getting there," their mother said. "Why don't you look out the window for a while? Look for the mileage markers or something."

Axel soon saw one, below a sign saying that it was 91 miles to the beach. He started his stopwatch and stopped it at the following mile marker. Seventy-two seconds had passed.

"Seriously, Dad, how long will it be?" Darius said.

"About an hour and a half."

"About? More than that or less?" Darius asked.

"Well, the traffic is pretty steady, so we should be able to keep this same speed all the way there. You guys figure it out," their father said.

A few moments later, Axel said, "We might as well relax. It's going to be longer than an hour and a half."

"What makes you say that?" Darius asked.

"I used my watch and the mileage markers," Axel said. "It took us 72 seconds to get from one mile marker to the next. I started the timer at the marker where it was 91 miles to the beach. That meant we had 90 miles to go when I stopped it."

Axel took out a pencil and paper. He said, "If it takes 72 seconds to go one mile, the length of time, in seconds, to go 90 miles would be 72 times 90—that's 6,480. Since there are 60 seconds in a minute, to find the number of minutes divide 6,480 by 60—that's 108 minutes. And since there are 60 minutes in an hour, 108 minus 60 leaves 48. That means it will take one hour, 48 minutes."

Cold Blooded Calculations

Henry was so excited that his parents had finally allowed him to get a pet iguana. They went to the pet store that day, and when they got there he ran straight to the tanks.

Henry had already picked out a medium-sized iguana, and he had things for the iguana to climb on. So he just needed a tank. Okay, easy enough, he thought.

He wanted to give it as much room to climb around as he could. There were three sizes of tanks. One had a base of 16 by 24 inches and was 12 inches high. The other had a base 16 by 16 inches and was 20 inches high. Another had a base of 20 by 20 inches and was 12 inches high. The cost of each tank was about the same.

Henry looked at the tanks for only a moment. "I'll take this one. It has the most space," he said, pointing to one of them.

"How did you figure that out so fast?" his father asked.

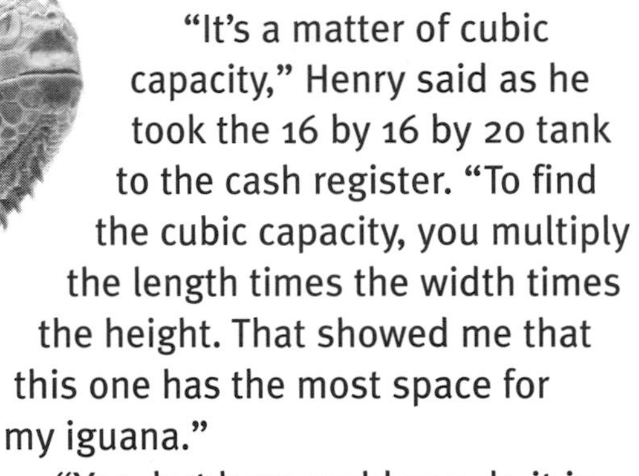

"It's a matter of cubic capacity," Henry said as he took the 16 by 16 by 20 tank to the cash register. "To find the cubic capacity, you multiply the length times the width times the height. That showed me that this one has the most space for my iguana."

"Yes, but how could you do it in your head so fast?" his father asked. "I think I'd need a pencil and paper for that."

"To get the exact number, yes," Henry said. "But we only needed to compare, so I simplified the calculations. Each of the dimensions was divisible by four. With the 16 by 24 by 12 tank, if you divide each number by 4 you're left with 4 by 6 by 3—72. The 20 by 20 by 12 tank becomes 5 by 5 by 3—75. The 16 by 16 by 20 tank becomes 4 by 4 by 5—80. That's the biggest of the three."

As they drove home, Henry got out a pencil and paper to check the exact figures. "The 20 by 20 by 12 tank is 4,800 cubic inches," he said. "The 16 by 24 by 12 tank is 4,608 cubic inches. The 16 by 16 by 20 tank is 5,120 cubic inches. That is the biggest, and it should give my new iguana plenty of room."

Ups and Downs

"Over the river and through the woods, to grandmother's house we go!" Jack and Courtney sang.

They were at a gas station near their house, filling the car for a trip from the mountains where they lived to their grandparents' summerhouse on the beach.

Hours later, just as they reached the house, a warning light came on that they were almost out of gas.

"That's pretty good mileage," their father said. "This car has a 15 gallon tank, and we used almost all of it to go about 500 miles. That's about 33 miles to the gallon. We usually only get about 25 miles per gallon, driving on highways like these. So that's about one-third better than usual."

"Should we get more gas now?" their mother asked.

"No, we won't be driving anymore until the day we leave. We'll get it then," he said.

They enjoyed their visit for the next few days and when it was over, they stopped to fill the tank for the drive home. They would be going back the same way they came.

"On the way back, we'll be going uphill from the beach to the mountains, and the car will have to work harder, so we'll use more gas," said Courtney. "Since we got gas mileage that was one-third better on the way down, we'll need gas again when we get two-thirds of the way home."

"Are you sure that's right?" Jack asked.

"You're assuming that because we got a third better gas mileage on the way down, we'll get a third less on the way back," Jack said.

"Right. Since we needed a full tank to get here, the tank will be just about empty when we're two-thirds of the way home," Courtney said.

"You're using the wrong starting point. The one-third difference is not from the 33 miles per gallon we got on the way here, it's from the usual mileage of 25 miles per gallon," Jack said. "One third of 25 is about 8, so we'll get about 17 miles per gallon on the way back home. Since the tank holds 15 gallons, we'll be able to go 255 miles before running out of gas—17 times 15. So we'll have to get gas again about halfway home, not two-thirds of the way."

51 Yuck Around the Clock

Mrs. Santos's class had just gotten new materials. There were pictures, maps and a desk clock that would replace the old wall clock that never worked right. However, when Mrs. Santos took the old clock down, there was a dark stain on the wall where it had been.

"Ew! When was the last time anybody cleaned back there?" Chloe asked.

"That's what happens over time," Mrs. Santos said. She took a cleaning cloth to the stain, but it was too dark to come out. "I'll ask the maintenance people to paint there, but it will be some time until they get to it," she said.

"How about if we cover up the spot with something else in the meantime?" Aiden suggested. He measured the spot; it was 10 inches across.

"How about this?" Connor said, measuring a picture. "It's an 11 by 11 square."

"Here's something bigger," Maria suggested, measuring another picture. "It's 8 inches tall by 20 inches long."

Aiden said, "Okay, now we figure out the—"

Chloe interrupted him. "I think the choice is obvious, don't you?" she asked.

"Maria, your rectangular picture has more square inches, but remember, we're trying to cover a round stain that's 10 inches in diameter," Chloe said. "That means whatever we use to cover it needs to be at least 10 inches in every direction. A picture that's only 8 inches in one direction would let some of the stain show. So even though the square picture has fewer total square inches than the rectangular picture, 121 versus 160, the square would cover the stain where the rectangle wouldn't."

Mixing It Up

"We will see you all next week at our next meeting!" said Mrs. Jackson. She was the leader of a mother-daughter book club that met at the library. "Wait, one more thing. We need someone to bring drinks. Anybody?"

"I will," Denise said. "I can bring punch."

"You're always bringing things. Let me help you," Mary said as they were walking out. "My mother has a great recipe for punch. It's half seltzer and half orange juice."

"Alright," Denise said. "I'll bring the juice, and you bring the seltzer."

At the next meeting, Mary left a two-liter bottle of seltzer on the table and noticed that Denise had brought in a two-quart container of orange juice. Denise emptied both ingredients into the punch bowl as Mary helped set up the chairs.

Denise brought Mary a cup of the punch. "I really like the way your mother's punch tastes," Denise said.

"I'm sure I'll like it too," Mary said, taking the cup. "But it won't taste quite the same."

"Why not?" Denise asked. "We followed the recipe."

"The ingredients were the same, but the amounts weren't," Mary said. "The recipe calls for half seltzer and half juice. We had two liters of seltzer and two quarts of juice. Liters and quarts are close in size, but they're not the same. A quart is 32 fluid ounces, while a liter is 33.8 ounces. That's 1.8 ounces more in each liter. Multiply that by two, because there are two liters, and that means there are 3.6 ounces more seltzer than juice."

Mary tasted it. "Actually, I like it better this way than the way my mom makes it. With more seltzer, the punch has more punch."

String Theory

Ms. Leonardo's class was having a contest. The students were broken into two teams, the left side of the classroom against the right. Each team was given a large box of the same size.

"I am giving each group a 48-inch piece of string," she said. "There is enough space in your boxes for you to lay out your string. There is only one catch. The two ends of your string must be touching. After you lay out your string, I'm going to drop one of these identical balls in each box. If the ball stops inside your string, everyone on that team gets two free homework passes. I will give you five minutes to lay your string out."

"What do we need five minutes for?" Alyssa said to her teammates, "It's completely random who's going to win."

"It might seem random, Alyssa, but soon you'll see why it isn't," their teacher said.

Alyssa's team finally settled on making a square with their string, 12 inches on each side.

When both teams had put their string down, Ms. Leonardo dropped the balls. The ball for the other team settled inside the string, but the ball for Alyssa's team didn't.

"We win!" yelled Jacob, who was on the other team.

"No fair!" said Alyssa. "Ms. Leonardo, their team cheated. Their string must have been longer than ours.

"No it's not," Jacob said. "We just figured out a way to make our string enclose more area."

"How?" asked Alyssa.

"You put your 48 inches of string into a square. That means your string enclosed an area of 12 times 12, which is 144 square inches," explained Jacob.

"We made our string into a circle," Jacob said. "The circumference was 48 inches, the length of the string. To figure out the area of our circle, we first had to know the radius. The circumference of a circle is two times pi times the radius. So to get the radius, we divided 48 by two—making 24—and then by pi. Pi is a little more than 3, so dividing 24 by that means the radius is a little less than 8—we rounded it down to 7. The area of a circle is the radius squared times pi. The square of seven is 49—7 times 7. We rounded pi down to 3 and multiplied it by 49, making 147 square inches. So even with just estimating like that and rounding down each time, we knew a circle would cover more area than a square."

"And if you do the math exactly, you'll see the difference was even larger," Ms. Leonardo said, going to the whiteboard. "Divide 48 by two to get 24 and then divide 24 by 3.14—using that as a more precise value of pi—means the radius was 7.64 inches. The square of that is 58.36, which, multiplied by 3.14, gives an area of 183.28 square inches. That's nearly 40 more square inches than the square covers. That increased the probability that the ball would stop inside their string, and as it turned out, it did."

54 Product Placement

Max had offered to stay after school to help with the fundraiser.

The parents' association was buying new supplies for the three rooms in his grade, and they had divided the cost among all the students. Each student in the three rooms needed to bring in $15.63. Mr. McGovern's room had 23 students, Mrs. Chang's room had 25, and Mrs. Bittle's room had 24.

The first room to bring in all of its money would get three days off from homework, the second two days off, and the third one day off.

Mr. Howard, the father of Max's classmate Daniel, was in charge of announcing the winner.

Each room had met its goal. Unfortunately, the envelopes were not marked with the names of the teachers, just the amounts in them.

"I know the white one came in first, the yellow one was second, and the brown one was third," Mr. Howard said.

Max looked at the envelopes. Mr. Howard's handwriting was so bad that Max couldn't make out the exact figures. The white envelope was some dollar amount and 75 cents. This yellow one was something and 49 cents, and the brown one was something and 12 cents.

Daniel glanced at the envelopes too. "So that tells us what we need to know," he said.

Max protested, "How could you figure that out in just a few seconds? I can't even make out the entire numbers!"

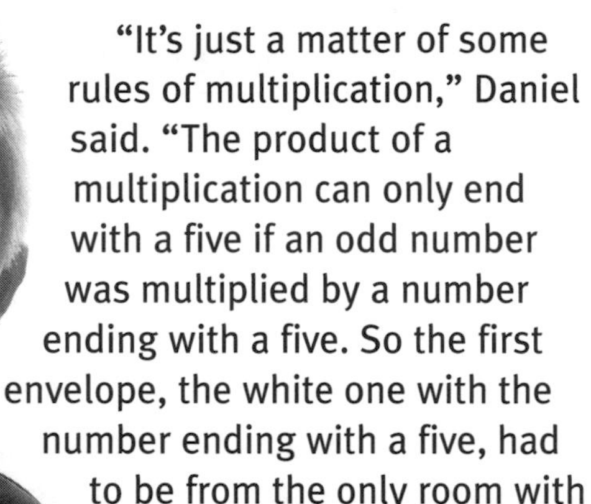

"It's just a matter of some rules of multiplication," Daniel said. "The product of a multiplication can only end with a five if an odd number was multiplied by a number ending with a five. So the first envelope, the white one with the number ending with a five, had to be from the only room with a number of students ending in a five—the 25 students in Mrs. Chang's class."

Daniel continued, "Of the other two numbers, one was odd and one was even. The product of a multiplication is odd only when you multiply two odd numbers. An even number times either an even or an odd number gives you an even number. So the odd number on the yellow envelope, the one that came in second, had to be the product of multiplying two odd numbers. So Mr. McGovern's room with the 23 students had to be in second place, and Mrs. Bittle's room with an even number of students, 24, is in third."

"The actual numbers," Mr. Howard said, looking at the envelopes, "are $390.75 for Mrs. Chang's room, $359.49 for Mr. McGovern's room and $375.12 for Mrs. Bittle's room. I guess I have to work on my handwriting."

Coupon Rate

Arianna was at the mall doing some shopping with Haley's family. Arianna had brought along $20, which she'd saved from her allowance so she could buy her sister a sweater as a birthday present. That morning she'd seen one advertised in the newspaper for $19.95. The newspaper also had a $1 off coupon, which she'd cut out.

They found the sweaters soon enough, but Arianna realized that she'd forgotten there was a sales tax of 5.5 percent. She was worried that she wouldn't have enough money.

"Can I borrow some change?" she asked Haley as they stood in line. "I'll pay you back when I get home."

"Sure, but can I look at that coupon first?" Haley replied.

"Okay, but what good will that do?" Arianna asked.

"It's a question of whether they take the value of the coupon off the price before they charge the taxes or after," Haley said. "That does make a difference."

She opened the calculator application on her cell phone. "To find how much 5.5 percent tax adds to $19.95, multiply 19.95 times 1.055. That comes to $21.04725, which we'll round up to $21.05. So if you subtract a $1 coupon off that, you'll need another nickel.

"But let's say they take the value of the coupon off first. Now you're paying tax only on $18.95. Multiplying 18.95 times 1.055 is $19.99225. Even if the store rounds up, your $20 would be enough."

As she looked at the coupon again, Arianna was happy to see that the taxes were charged after the value of the coupon was deducted. She didn't have to borrow anything from Haley after all.

56 Turning Up the Volume

"Now, class, don't lose your marbles over this assignment," Mr. Vaughan said, laughing as he handed each of them a bag of marbles.

"Your task over the weekend is to find the volume of one of these marbles in cubic centimeters," he continued. "I understand that there might be slight differences in size among them. What I'm looking for is that you understand the concept of the volume of a . . . what shape is a marble?" he asked, to see if the students were listening.

"A sphere," they answered.

"And by the way, I need those marbles back Monday morning when you turn in your answers," he said. "I don't want to see any marbles tournaments going on around here."

Aaron, Travis and Katelyn turned in their papers on Monday and got them back the next day. They all had gotten a check mark indicating that their answer was acceptable.

"That was tough, measuring those little marbles with a tape measure," Katelyn said as they walked out of class.

Aaron agreed. "I'm glad he said he was more interested in showing we knew how to do the calculation than in getting an exact measurement," he said.

"I didn't even use a tape measure, and he wrote on my paper that I was the most accurate of everyone," Travis said.

"How could that be?" Katelyn said.

"I'll tell you, but first tell me how you got your answers," Travis said.

Katelyn said, "Well, I used a tape measure to find the circumference of one marble in centimeters. Since the circumference of a circle is two times pi times the radius, to find the radius I divided the circumference by two and then by pi."

"I did the same thing," Aaron said. "Once I had the radius, I could calculate the volume: the volume of a sphere is four-thirds times pi times the cube of the radius."

"I know those formulas too," Travis said, "but I noticed he stressed that this was about the concept of volume. Also, he gave each of us a whole bag of marbles when one marble would have been enough to measure volume the way you guys did. So I went to the after-school period in the science room, got the largest graduated cylinder they have, the 1,000 milliliter one, and put in as many marbles as would fit below the 1,000 milliliter line. I filled it up to the 1,000 milliliter line with water and then poured the water into another graduated cylinder, to see how much water I'd put in the first one. I subtracted that number from 1,000 to find the volume of the marbles in milliliters—that is, how much room they took up in the first cylinder. I divided the result by the number of marbles I had put in, giving me the average volume of each marble. Since a milliliter is equal to a cubic centimeter, I had my answer."

57 Down to the Last Drop

"I think we have a problem here," Isabella said to her mother.

Her mother gave her a look, as if to say she didn't need any more problems just now.

They were setting up the kindergarten room where Isabella's younger sister Grace went to class. It was the 100th day of the school year, and the school always set out a snack when the kids came to class that day.

It was just before school was supposed to start. Isabella was in a hurry to get to her middle school on time, and her mother needed to get to her job. Other parents had dropped off juice containers, cookies, plates and cups, and then rushed off.

"What's wrong?" her mother asked.

"Well, there are 26 kids in the class, and we have a one-gallon container of juice plus a one-quart container. We have two dozen 8-ounce cups and one dozen 6-ounce cups," Isabella said. "I guess we'll just have to give some of them more juice than others."

"You obviously haven't ever had to manage a group of kindergarteners," her mother said. "If everything isn't exactly the same, there will be trouble."

"I have an idea," Isabella said.

"I'm all ears," her mother said

"We'll use the 6-ounce cups to measure juice into the 8-ounce cups," Isabella said. "One gallon plus one quart is 160 ounces—128 plus 32. If we gave everyone 8 ounces, we'd only have enough for 20 kids—160 divided by 8 is 20. To find out how much we can give to have an equal amount for everyone, we have to divide 160 by 26. That's just above 6 ounces per child. So we'll set out all the 8-ounce cups but use one of the 6-ounce cups to measure how much to put in each. Once we use up the 24 8-ounce cups, we'll fill two of the 6-ounce cups all the way."

58 A Fan of Keeping Cool

"I can't believe that house doesn't have air conditioning," Nicholas said.

"Well, this is Maine. It usually doesn't get this hot, so the houses don't need it," his father said.

Nicholas, his sister Sarah and their parents were in an appliance store. They were vacationing for a week in a rental house near a lake in Maine. But they had arrived during a heat wave that was supposed to last all week. None of them could sleep the night before because it was too hot, so they decided to buy a large window fan to draw air through the house. The store had two kinds of fans. Both were the same size and about the same cost.

"Let's get the one that will move the most air," said their mother.

"This fan has five different speeds," Nicholas said. "The lowest speed is 200 rpm. What's rpm?"

"Revolutions per minute," their father said. "In other words, how fast the fan blade is turning."

"Okay, so the lowest speed of this fan is 200 rpm, and it says that each of the four faster speeds is twice as fast as the previous one," Nicholas said. "That means the second speed would be 400 rpm, the third 600, the fourth 800 and the top speed is 1,000 rpm. This other fan has only three speeds, but its fastest speed is 2,400 rpm, so we should buy the three-speed one, don't you think?"

"No, we should buy the five-speed fan," Sarah said. "Remember, that fan's speed doubles with each faster setting. That means you need to double the previous speed each time, not keep adding the amount of the first increase to the starting speed. Since the first speed is 200 rpm, the second would be 400, the third would be 800, the fourth 1,600, and the fifth 3,200 rpm. So the five-speed fan's top speed is faster."

Overdue Blues

“Hey, Martha, have you seen that mystery book you took out from the school library?” her mother asked. “We got a letter from the librarian today saying you’ve had it for 80 days and haven’t returned it. You were supposed to take it back after two weeks.”

“Oops. I think I lost it somewhere,” Martha said. “How can I make it up to the school?”

“Well, the note says we can either return the book and pay the fine or pay them enough to buy a replacement book,” her mother said.

“How much is each?” Martha asked.

“It says the fine is 10 cents a day for every day a book is overdue. And it says the book you borrowed costs $7.50,” her mother said.

The next day Martha went into the library. The librarian asked her, “What are you going to do about the overdue book?”

"I'm going to keep looking for it," Martha said. "Yesterday was the 80th day since I borrowed it, so today is the 81st day. But since I could keep it for two weeks, 14 days, without a fine, it's 67 days overdue, not 81 days overdue. At 10 cents a day, that's $6.70. So I have another week to find it and still spend less than the $7.50 cost to replace it."

"Good luck! I hope you find it in time," said the librarian.

Paper Chase

"What do they use this thing for, anyway—to weigh trucks?" Diego asked his lab partner Jenna.

The science lab had several good scales, including a triple-beam scale that measured with accuracy down to 0.1 grams. But their teacher, Mr. Holmes, had gotten out the old scales for this assignment. He said he was going to order new shelves for the storage room and needed to know how strong they had to be to hold the packages of paper he kept there. Each package was 500 pages. He wanted to know exactly how much a sheet weighed.

The students thought that would be easy enough to figure out, until Mr. Holmes pulled out the old scales. They could hold a lot of weight but were accurate only to 10 grams.

Jenna laid a sheet of paper on the scale. It didn't budge. She tried another. This time the balance moved.

"So two sheets is about 10 grams, meaning one sheet is 5 grams. More or less," Diego said.

"But he's not looking for a 'more or less' answer," Jenna said.

"Well, then, there's only one thing we can do," he said.

"What's that?" asked Jenna.

"Let's get some more paper," Diego said. They went into the storage room and returned with two full packages of 500 sheets each, unwrapped the packages and put the paper on the scale, after removing the original two sheets. The weight came to 5,440 grams.

"Since 1,000 sheets weigh 5,440 grams, the weight of each sheet is one-thousandth of that. To divide by 1,000, all we have to do is move the decimal point three places to the left. So one sheet is 5.44 grams," Diego said.

"I see—just because the scale is imprecise doesn't mean we have to be," Jenna said.

BONUS SECTION

Five More Minutes of Mysterious Math

1 Ice Cream Anyone?

"Welcome to Cora's Ice Cream Parlor," a lady said from behind the counter.

She was glad to fill up seats in her ice cream shop on a chilly day, but she hadn't counted on all these energetic middle school girls celebrating the end of their fall field hockey season.

"Sir, you phoned earlier to reserve 21 seats for your team, right?" she said to the coach, Mr. Lee. "We're all set up for you," she said, pointing to tables that had been set with spoons and empty bowls.

He had told the girls he would treat them to two scoops of ice cream each.

"Coach Lee," said Claire, one of the players. "You're always saying that each one of us is unique, aren't you?"

"Yes," he said slowly, not sure he wanted to know what was coming next.

"So, every one of us wants something different from anyone else," she said, and all the other girls started laughing.

"Boy, they're a picky bunch, aren't they?" Cora said. "I don't even have 21 different flavors, I only have 12. What can I do?"

"It's a matter of how many combinations are possible. To find that, you add the number of samples to each number below it," Claire said to Cora.

"I'm not following you," Cora said.

"For example, if you had three flavors, that would be 3 plus 2 plus 1, making 6 possible combinations," she said. "Say you have vanilla, chocolate and strawberry. To use all the possible combinations of two scoops, you'd have one bowl with vanilla and chocolate, a second with chocolate and strawberry, a third with vanilla and strawberry, and three more bowls that have two scoops of the same flavor. That's 6 different combinations for three samples. It works the same way with each additional sample flavor you add. With four flavors, you'd have 10 possible combinations—4 plus 3 plus 2 plus 1. With five flavors, it is 15 possible combinations—5 plus 4 plus 3 plus 2 plus 1. To get 21 possible combinations, you actually only need six flavors—6 plus 5 plus 4 plus 3 plus 2 plus 1 is 21."

"For that explanation, young lady," said Cora, "you get the first choice of flavors."

2 Puttin' on the Hits

"A cassette player!" James said. "That's all we have for tunes?"

It was afternoon break time on the first day of their two-week stay at summer camp. Campers had two hours of free time, but they couldn't bring any personal electronic devices or other valuables to camp. That left an old cassette player and some tapes that looked like they had been there for many years as the only source of music for the cabin.

"My dad has one of these," Anthony said.

"How do you work it?" asked James.

Anthony picked out a tape labeled Groovy, Man! One Hour of the Best Music of the Sixties. "Well, you put the tape in this slot and push the play button. These numbers"—he pointed to a counter reading 145—"increase as the tape plays. When the tape hits the end, you flip it over and play the other side."

They listened to the entire tape from the beginning, really liking some of the songs. When they had played both sides, James noticed the counter read 865.

"That 'Hey, Jude' song must be the longest song ever," James said.

"I bet that one called 'Innagaddavida' or something was even longer," Anthony said.

"Well, if we had something that tells time, we could play the songs again and time them," James said. "But since we don't, how will we ever know?"

"We'll use the tape counter," Anthony said. "The tape played for 60 minutes and the counter went from 145 to 865. That means that in 60 minutes, the counter increased by 720—865 minus 145. Dividing 720 by 60—which is the same as dividing 72 by 6—means that the counter advanced 12 numbers per minute, or 1 number every 5 seconds."

"I get it," James nodded. "So we find the place on the tape where each song starts, see how much the counter advances while the song is playing, and we'll be accurate within 5 seconds, which should be close enough."

3 And They Call This a Fair?

Mrs. Grabowski's class was working at the school fair to help raise money for new supplies for the math room. Kendall and her best friend Ruby decided to make a game with 20 rectangles of flat cardboard that were three inches long and two inches wide.

The sign on their table said:

Win a prize by proving you know which way these rectangles can be arranged to cover the most area.

Two boys they knew, Micah and Sean, tried five times. Mrs. Grabowski, the judge, rejected all the different arrangements they made.

"This game is impossible," Micah said.

"Maybe to you, Micah, but it really is possible," Ruby said.

After the fair was over and nobody had won, Sean came up to Ruby and Kendall.

"Okay, what's the correct answer?" he asked.

"Remember, the goal was to prove you know which way they cover the most area," Kendall said. "They cover the same area no matter how you arrange them. All you had to do was say that."

4 Cold as Ice

"Now we have to pour the salt into the water." Mr. Kaufman was leading his class in an experiment to see whether water would freeze at a different temperature if salt was added to it.

"Good, good!" he said when the water finally froze. "So we have now proven by experimentation that putting salt in water makes it freeze at a lower temperature—28 degrees, in our trials. Now don't forget that you have to write a lab report on what we've done, and it is due tomorrow!"

The class groaned as they packed up their books to go. In the hall, Julian ran up to Katherine and asked her what temperature scale they were supposed to write their lab report in. "I'd say we should to write it in Celsius," she said. "That's more commonly used in science than Fahrenheit."

"Okay, thanks Katherine," he said.

When Julian got home, he wrote up his paper describing how they had made water freeze at 28 degrees Celsius.

The next day in class, Mr. Kaufman asked Julian for his paper. At the first glance, he said. "Okay Julian, see me at lunch."

"Uh-oh," Julian thought. "What does he want to say to me?"

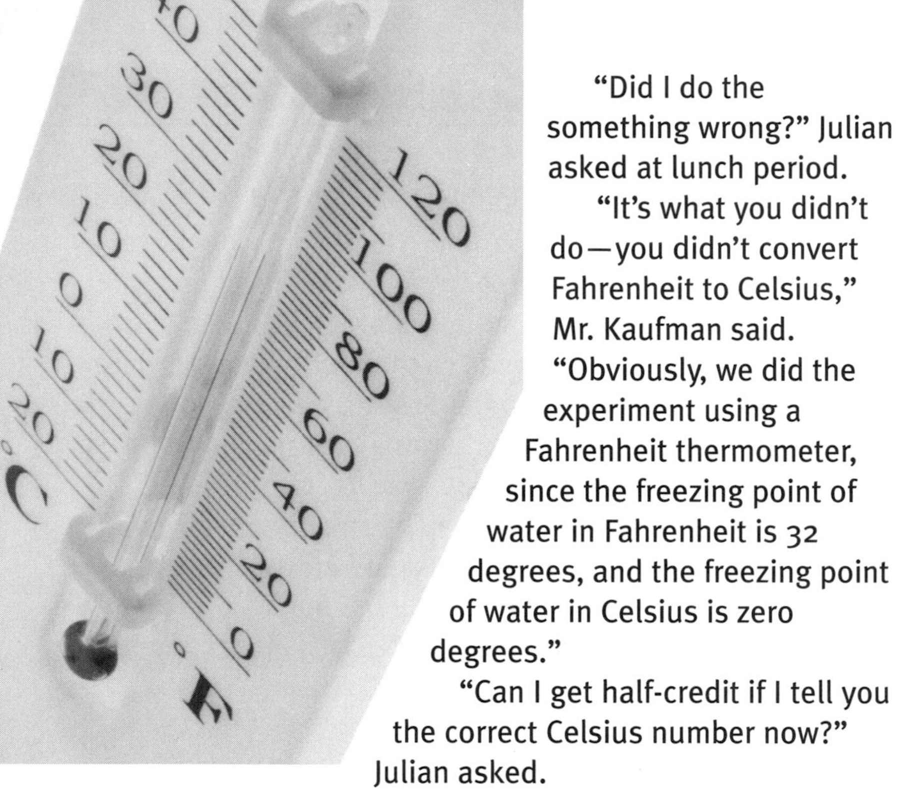

"Did I do the something wrong?" Julian asked at lunch period.

"It's what you didn't do—you didn't convert Fahrenheit to Celsius," Mr. Kaufman said. "Obviously, we did the experiment using a Fahrenheit thermometer, since the freezing point of water in Fahrenheit is 32 degrees, and the freezing point of water in Celsius is zero degrees."

"Can I get half-credit if I tell you the correct Celsius number now?" Julian asked.

"Okay, if you tell me how to do the conversion," Mr. Kaufman said.

"First you subtract 32 from the Fahrenheit temperature, then you multiply by 5/9," Julian said. "28 minus 32 is -4, times 5 is -20. Divide that by 9 and it's negative 2.22. So the water froze at -2.22 degrees Celsius."

"Good job, Julian," said Mr. Kaufman. "I knew you knew how to do it. You just have to be more careful next time and make sure you do the conversions."

5 A Switch in Time

Brady spent a week each summer visiting his grandparents. One day in 2009 he went with his grandmother to a flea market.

There were rows of tables with books, record players, musical instruments, appliances, toys, dolls—just about everything.

"Sometimes you can find things that are worth a lot more than you pay," his grandmother said, giving him $20. "Spend this wisely."

He spent it on a box of old comic books that he was sure were worth more. There was a second box that also had some rare comic books but he didn't have enough money to buy it.

He found his grandmother at a table full of old silverware, plates, cups and other tableware. She loved history. She already bought an old book titled The Civil War: 1861-1865. She turned over a dish and read the inscription on the back: "Manufactured in Richmond, Virginia, USA."

"These old Southern serving dishes are collector's items," she said. "Only $30. I think I'll buy this."

"You really ought to," the man selling them said. "These dishes are genuine antiques—exactly 145 years old."

Brady took her aside and said, "I think you should buy that other box of comic books instead."

"Why?" she asked.

Brady laid the dish back on the table and led her away. He said, "This is 2009, so 145 years ago was 1864. In 1864, Richmond was the capital of the Confederacy. They didn't consider themselves part of the United States of America. They would have stamped it as CSA, for Confederate States of America, not USA. So that dish has to be a fake, and buying it would be a waste of money. You can really help yourself by doing a little math."

Discover
One Minute Mysteries:
65 Short Mysteries You Solve With Science!

Pumpkin Patch

"Where's Linus? We have the Great Pumpkin," Belinda said.

Madison laughed. It was Halloween, and all the neighbors around the cul-de-sac were going in on a contest to see who carved the best pumpkin. The prize was a gift certificate to the video store.

She and Belinda had carved a pumpkin with the face of a witch and then used colored paper to cover the openings so that the light from the candle inside would make the face glow green. They lit the candle, put the top back on the pumpkin and went out trick-or-treating

As they went from house to house, they looked over the other entries for the contest. Carly had done a nice carving of a ghost, and Peter had used an odd-shaped pumpkin to make a funny-looking frog. They came to Sam's house just as he was setting out his pumpkin.

"This is the winner right here," Sam said, in a not-too-friendly voice. Madison had to admit, he'd done a good job of carving a haunted house, but she still knew her and Belinda's pumpkin was better.

"Watch out for him," Belinda said as they went to the next house. "He'll probably smash the other pumpkins just so he can win."

After they made a lap of the neighborhood, they returned to Madison's house to drop off the candy before everyone would meet for the winner to be announced. Their pumpkin was not smashed, but the candle was out, so it didn't look like anything special.

"I know who's to blame," said Madison.

"Who?" Belinda asked.

"It's our fault," Madison said. "The candle went out on its own. When we covered up the holes with the paper and put the lid on, we cut off the oxygen supply. Fire needs oxygen to keep going. Looks like Sam's going to be the winner. Maybe he'll invite us to watch the movie he gets with that gift certificate."

2 Stuck With the Mud

At Grady and Jim's school, there was a courtyard enclosed by classrooms on all sides. There were flower beds, berry bushes, trees with bird houses and even a little pond with lily pads, frogs and turtles. A couple of times a year, there was a workday during each science class to fertilize the plants, cut off dead branches and otherwise keep the courtyard in shape. This was one of those days, a cloudy, cool, damp day in the fall.

Jim and Grady had just finished tying back some branches with strong string when the teacher, Mr. Burke, assigned them another chore. He handed them a cloth bag and a shovel.

"Dig a little soil from the edge of the pond, dry it out, and bring it into the classroom. We're going to test it for bacteria in a few minutes," he said.

"Dry it out?" Grady asked.

"Damp is okay. Just dry enough so it doesn't drip. The custodian will be upset if we get mud on the floor," Mr. Burke said as he left them.

Grady held the bag while Jim shoveled in some gooey dirt, then laid the bag on the ground. Some water oozed out, but what was left was still very wet.

"Man, it will take forever for this to dry out. And it's almost time to go back inside," Grady said.

"All we have to do is speed up the process," Jim said.

"How?" Grady asked. "If you expect me to squeeze that glop out, you're nuts."

"Stand back," Jim said. He used the string to tie the cloth bag shut and to swing the bag in circles. Water came flying out of the bag.

"It's just a matter of using centrifugal force," Jim said. "When the bag is spun in a circle, the water is free to move through the bag while the dirt is held inside and is pressed against the inside of the bag. So I knew it would squeeze the water out, leaving only damp dirt inside.

"By the way," he added, "centrifugal force is what's called a 'pseudo' force, compared with 'real' forces like gravity or magnetism that have external causes. Centrifugal force exists here because the string is keeping the bag from continuing in a straight line, which inertia otherwise would make the bag follow once it was set in motion. The string is causing the bag and the dirt inside to constantly change direction by spinning the bag in a circle. The bag doesn't restrict the water, so the inertia of the water – which keeps it moving in one direction – makes the water fly out of the bag."

Freeze Fall

It was late winter, and the temperature had just fallen after several mild days. To make the walk home from school even colder, it had rained earlier, and a chilly mist still hung in the air.

Tom and Evan glanced up at the flashing clock in front of the bank. It said "32°F, 0°C."

They stopped in a candy store for a snack and to warm up before they continued on their way home.

Their shoes splashed through puddles as they headed toward the railroad bridge. The bridge, several hundred yards long, had a narrow sidewalk next to train tracks, where the tracks crossed the river far below. It could get scary crossing the bridge when a train was on it. But the only way to avoid it was to take a different route that added ten minutes to the walk.

"I think we should go the long way," Evan said. "The bridge is probably icy."

"We haven't seen any ice. These sidewalks are just wet," Tom said.

A few moments later they were on the bridge. Tom's foot slipped on a patch of ice and he fell.

"I told you so," Evan said, teasing him.

"How did you know there would be ice here when there isn't ice anywhere else?" Tom asked as Evan helped him up.

"The Earth absorbs heat from the Sun and radiates that heat back out. Up until the bridge, there is ground under the sidewalks. The ground provides some insulation and keeps the sidewalks above the freezing point, even though the air temperature itself is at the freezing point," Evan said. "But underneath the bridge there is just cold air without any insulation, so the surface on the bridge freezes first."

Left In the Dark

“You’re kidding,” Dejon said.

“Nope,” Damien said. “My parents told me this morning. The power’s been out in the school almost the entire break.”

Dejon, Matt, Quincy and Damien were walking to school on the first day back after the winter break of more than a week.

“What happened?” Matt asked.

“Dad said they were doing some kind of work on the electrical system and they had to shut down the power the whole time,” Damien said. “He said it’s a good thing the weather has been so mild, so at least the pipes didn’t freeze and break.”

As they approached the school, they could see that the lights were on. Everything seemed okay. But then Quincy realized something. “Our experiments!” he said.

For several weeks before the break, they had been doing projects that measured the effects of light on growth, using terrariums with lights on automatic timers. Dejon had been growing bean sprouts, Matt had been growing mushrooms, Quincy had been growing cucumbers and Damien had been growing corn. It was going to be a big part of their grade for that marking period.

“What about the experiments?” Dejon asked.

“We’ll have to start all over again,” Quincy said. “Without those automatic lights, they’ll be ruined.”

“Not all of them,” Damien said. “I can think of one of them that should be all right with very little light.”

“Whose?” Quincy asked.

“You’re in luck, Matt,” Damien said. “Mushrooms don’t need much light to grow. That’s why you mostly find them in shady places.”

5 Occupational Hazards

It was the start of Career Week in science class. The students had to pick an area of science they found interesting and then research what it would be like to work in that job.

"J.L., let's start with you," said their teacher, Mr. Chisek. "What career are you going to study?"

"I've always liked space," J.L. said. "I think meteors are really neat. So I'm going to look into being a meteorologist."

"I like space, too," George said. "I just love looking at the stars. So I'll research being an astrologer."

Mr. Chisek said, "How about something a little more down to Earth?"

"I'm interested in plants," Jahari said. "I'm going to do my report on how to be a botanist."

"Is there anyone interested in rocks and minerals?" the teacher asked.

"Me," Trisha said. "I'll do mine on what it's like to be a geographer."

"Anyone interested in electronics?" Mr. Chisek asked.

Brandon raised his hand. "I love messing around with radios. I'm going to research radiology," he said.

Tucker leaned over and whispered to Acquan, "I'm surprised Mr. Chisek isn't saying anything. But I guess he's going to let them find out the hard way. All of them except for one are in for a surprise."

"I know what you mean," Acquan said. "Meteorology is not the study of meteors, it's the study of weather. And astrology isn't the study of the universe, astronomy is. Astrology isn't a science, it's a belief that the stars and planets affect our personalities and our lives."

Tucker said, "Trisha will soon find out that geology is the study of things found in the ground and that geography is the study of the Earth's physical features. And radiology isn't about radios, it's about using X-rays and radioactive substances to detect and treat disease. Jahari is the only one who's right; botany is about studying plants. Being accurate is really important in science!"

Science, Naturally! wishes to thank our wonderful team of Project Editors whose enthusiasm, eagle eyes and critical reading helped shape this into the fun and wonderful book that it is!

Jennifer Zoon, Kensington, MD
Danielle Donaldson, Norco, CA
Nora Goldman, Philadelphia, PA
Andrea W. Bailey, Washington, DC
Kristin Francoz, New York, NY
Casey Heilig, Washington, DC
Stine Bauer Dahlberg, Washington, DC
Molly Katharine Nelson, Washington, DC
Elaine Nicole Simeon, San Francisco, CA

Index

Photo and Illustration Credits

Page iii: Image by Michael Brown; Dreamstime.com
Page v: Image by Madartists; Dreamstime.com
Page 11: Photo by Patti Yoder
Page 13: Photo by Patti Yoder
Page 15: Illustration by Andrew Barthelmes, Peekskill, NY
Page 18: Photo by Patricia Hofmeester; Dreamstime.com
Page 20: Photo by jeancliclac; istockphoto.com
Page 22: Photo by Pete Saloutos; Dreamstime.com
Page 24: Photo by Evgeny Dontsov; Dreamstime.com
Page 26: Photo by luckynick; Dreamstime.com
Page 28: Photo by Nick Monu; istockphoto.com
Page 30: Photo by Ron Chapple Studios; Dreamstime.com
Page 32: Photo by Stanislaw Kostrakiewicz; Dreamstime.com
Page 34: Photo by Monkey Business Images; Dreamstime.com
Page 36: Photo by Andrejs Pidjass; Dreamstime.com
Page 38: Photo by Peter Chigmaroff; Dreamstime.com
Page 40: Photo by James Steidl; Dreamstime.com
Page 42: Image by Bruno1998; Dreamstime.com
Page 44: Photo by Ken Liu; Dreamstime.com
Page 46: Photo by Lev Olkha; Dreamstime.com
Page 47: Illustration by Andrew Barthelmes, Peekskill, NY
Page 50: Photo by Ivan Paunovic; Dreamstime.com
Page 52: Photo by Winzworks; Dreamstime.com
Page 54: Photo by Oleg Mitiukhin; Dreamstime.com
Page 56: Photo by Thomas Gordon; istockphoto.com
Page 58: Photo by Brad Sauter; Dreamstime.com
Page 60: Photo by Terry Alexander; Dreamstime.com
Page 62: Photo by Meg380; Dreamstime.com
Page 64: Photo by Charles Knox; Dreamstime.com
Page 66: Photo used by permission from Rocking Horse Ranch Resort
Page 68: Photo by Cstar Enterprises; istockphoto.com
Page 70: Photo by Cami Bruggner; Bigstockphoto.com
Page 72: Photo by Delmas Lehman; Dreamstime.com
Page 74: Photo by Robert Magorien; Dreamstime.com
Page 76: Photo by All-Kind-ZA Photography; istockphoto.com
Page 78: Photo by The Desktop Studio; istockphoto.com
Page 79: Illustration by Andrew Barthelmes, Peekskill, NY
Page 82: Photo by U Star PIX; istockphoto.com
Page 84: Photo by Thor Jorgen Udvang; Dreamstime.com
Page 86: Photo by Monkey Business Images; Dreamstime.com
Page 88: Photo by Rich Legg; istockphoto.com
Page 90: Photo by Jim Boardman; Dreamstime.com
Page 92: Photo by Kathy Wynn; Dreamstime.com
Page 94: Photo by Mwproductions; Dreamstime.com
Page 96: Photo by Pete Niesen; Dreamstime.com
Page 98: Photo by JSABBOTT; istockphoto.com
Page 100: Photo by Elysabeth McReynolds; Bigstockphoto.com
Page 102: Photo by Galina Barskaya; Dreamstime.com
Page 104: Photo by Ron Chapple Studios; Dreamstime.com
Page 106: Photo by Elena Elisseeva; Dreamstime.com
Page 108: Photo by ktmoffitt; istockphoto.com
Page 110: Photo by Susan Leggett; Dreamstime.com
Page 111: Illustration by Andrew Barthelmes, Peekskill, NY
Page 114: Image by Robodread; Dreamstime.com
Page 116: Photo by Gemphotography; Dreamstime.com
Page 118: Photo by Andres Rodriguez; Dreamstime.com
Page 120: Photo by istockphoto.com
Page 122: Phot by Elena Elisseeva; Dreamstime.com
Page 124: Photo by kiep; BigStockPhoto.com
Page 126: Photo by Gina Smith; Dreamstime.com
Page 128: Photo by Alexey Romanov; Dreamstime.com
Page 130: Phtoto by digital planet design; istockphoto.com
Page 132: Photo by Monkey Business Images; Dreamstime.com
Page 134: Photo by Doug Olsen; Dreamstime.com
Page 136: Photo by Monkey Business Images; Dreamstime.com
Page 138: Photo by Darko Despotovic; Dreamstime.com
Page 140: Photo by Koh Sze Kiat; Dreamstime.com
Page 142: Photo used by permission from Pennsylvania Scale
Page 143: Illustration by Andrew Barthelmes, Peekskill, NY
Page 146: Photo by Randy Harris; Dreamstime.com
Page 148: Photo by Dan Talson; Dreamstime.com
Page 150: Photo by Baloncici; Dreamstime.com
Page 152: Photo by Homestudiofoto; Dreamstime.com
Page 154: Photo by AndreasReh; Dreamstime.com
Page 155: Illustration by Andrew Barthelmes, Peekskill, NY
Page 158: Photo by Jafaris Mustafa; Dreamstime.com
Page 160: Photo by Grzegorz Wolczyk; Dreamstime.com
Page 162: Photo by Junkgirl; Dreamstime.com
Page 164: Photo by Marcos Moscardini; BigStockPhoto.com
Page 166: Photo by Tomasz Trojanowski; Dreamstime.com
Page 171: Photo by Patti Yoder

About the Authors

Eric Yoder is a writer and editor who has been published in a variety of magazines, newspapers, newsletters and online publications on science, government, law, business, sports and other topics. He has written, contributed to or edited numerous books, mainly in the areas of employee benefits and financial planning. A reporter at The Washington Post, he is also the award-winning coauthor, along with his daughter, Natalie, of the wildly popular *One Minute Mysteries: 65 Short Stories You Solve With Science!* He and his wife, Patti, have two daughters, Natalie and Valerie. He can be reached at Eric@ScienceNaturally.com.

Natalie Yoder is a high school student whose favorite subjects are science, photography and English. A sports enthusiast, she participates in gymnastics, field hockey, track, soccer and diving. She also enjoys reading, writing, playing the clarinet, playing with the family beagle, Trevor, and listening to music. She has appeared, along with her father, on National Public Radio to talk about their work on *One Minute Mysteries: 65 Short Stories You Solve With Science!* She loved helping to create and shape these math mysteries. She is hoping to work in marine biology after college. She can be reached at Natalie@ScienceNaturally.com

About *Science, Naturally!*

Science, Naturally! is committed to increasing science and math literacy by exploring and demystifying educational topics in fun and entertaining ways. Our mission is to produce products—for children and adults alike—filled with interesting facts, important insights and key connections. Our products are perfect for kids, parents, educators and anyone interested in gaining a better understanding of how science and math affect everyday life.

Our materials are designed to engage readers by using both fiction and nonfiction strategies to teach potentially intimidating topics. Some of the country's premier science organizations designate our materials among the best available and highly recommend them as supplemental resources for science teachers. In fact, **all** of our titles have been awarded the coveted "NSTA Recommends" designation.

You can listen to audio versions of our books, enjoy author interviews and explore our multimedia resources (including iPhone applications) at www.SoundbiteScience.com, your source for all our audio, electronic and mp3 selections.

For more information about our publications, to request a catalog, to be added to our mailing list, or to learn more about becoming a *Science, Naturally!* author, give us a call or visit us online.

Science, Naturally! books are distributed by National Book Network in the U.S. and abroad and by Mariposa Press in France.

Science, Naturally!®
725 8th Street, SE
Washington, DC 20003
202-465-4798
Toll-free: 1-866-SCI-9876
(1-866-724-9876)
Fax: 202-558-2132
Info@ScienceNaturally.com
www.ScienceNaturally.com

If you loved this book, don't miss...

One Minute Mysteries: 65 Short Mysteries You Solve With Science!

By Eric Yoder and Natalie Yoder

These short mysteries, each just one minute long, have a fun and intersting twist—you have to tap into your knowledge of science to solve them! Enjoy 65 brain twisters that challenge your knowledge of science in everyday life.

"Turns kids into scientists...stimulating and great fun for the whole family!" —Katrina L. Kelner, Ph.D., Science Magazine

"Parents and kids alike will be challenged...A great way to grow a young scientist—or improve an old one!"
—Julie Edmonds, Carnegie Academy for Science Education

As heard on National Public Radio

Audio renditions of selected mysteries can be heard at www.SoundbiteScience.com and www.REALscience.us.

Ages 8-12
ISBN: 978-0-9678020-1-5
Paperback: $9.95

101 Things Everyone Should Know About Science

By Dia L. Michels and Nathan Levy

"Succinct, cleverly written...should be on everyone's bookshelf!"
—Katrina L. Kelner, Ph.D. *Science* Magazine

"Readers will devour the book and be left eager for the 102nd thing to know!" —Margaret Kenda, *Science Wizardry for Kids*

"Their book challenges our understanding, intrigues us and leads us on a voyage of discovery!"
—April Holladay, author, *WonderQuest.com*

Why do you see lightning before you hear thunder? What keeps the planets orbiting around the Sun? Why do we put salt on roads when they are icy? What metal is a liquid at room temperature? And the burning question: Why do so many scientists wear white lab coats?

Science affects everything—yet so many of us wish we understood it better. Using an engaging question-and-answer format, key concepts in biology, chemistry, physics, earth science and general science are explored and demystified. Endorsed by science organizations and educators, this easy-to-tackle book is a powerful tool to assess and increase science literacy. Perfect for kids, parents, educators and anyone interested in gaining a better understanding of how science impacts everyday life.

Ages 8-12
ISBN: 978-0-9678020-5-3
Paperback $9.95

101 Things Everyone Should Know About Math

By Marc Zev, Kevin Segal, and Nathan Levy

"...kids of all ages will find this book engaging and challenging!"

—Carole Basile, Ed.D. Co-Director, Rocky Mountain Middle School Math and Science Partnership

"Whether you excel in or struggle with math, there's something in this book for you!"

—James Singagliese, Principal, Holy Family School

"A wonderful book for making complex topics approachable and helping readers discover the fascinating world of math!"

—Rachel Connelly, Ph.D. Economics, Bowdoin College

Math is a critical part of our everyday lives. The second title in the award-winning "101 Things Everyone Should Know" series helps you understand how you use math dozens of times—every day. With entertaining connections to sports, hobbies, science, food, and travel, mathematical concepts are simplified and explained using clear, real-life explanations. You'll even learn some fun trivia and math history! Using an engaging question and answer format, *101 Things Everyone Should Know About Math* is perfect for kids, parents, educators, and anyone interested in the difference between an Olympic event score of 9.0 and Richter scale score of 9.0.

Ages 10-14
ISBN: 978-0-9678020-3-9
Paperback $9.95

If My Mom Were a Platypus: Mammal Babies and Their Mothers

By Dia L. Michels • Illustrated by Andrew Barthelmes

"As engaging visually as it is verbally!"

—Dr. Ines Cifuentes, Carnegie Academy for Science Education

"The animal facts . . . are completely engrossing. Most readers are sure to be surprised by something they learn about these seemingly familiar animals."

—Carolyn Baile, *ForeWord* magazine

NSTA recommends™

NATIONAL SCIENCE TEACHERS ASSOCIATION

Middle grade students learn how 14 mammals are born, eat, sleep, learn and mature. The fascinating facts depict how mammal infants begin life dependent on their mothers and grow to be self-sufficient adults. This book highlights the topics of birth, growth, knowledge and eating for 13 different animals. All stories are told from the baby's point of view. The 14th and final species is a human infant with amazing similarities to the other stories. With stunning full color and black-and-white illustrations and concise information, this book helps children develop a keen sense of what makes mammals special.

Ages 8-12. Curriculum-based Activity Guide with dozens of fun, hands-on projects available free of charge at www.ScienceNaturally.com

ISBN: 978-1-930775-35-0	Hardback book	$16.95
ISBN: 978-1-930775-44-2	Hardback + 15" plush platypus	$29.95
ISBN: 978-1-930775-19-0	Paperback book	$ 9.95
ISBN: 978-1-930775-30-5	Paperback + 15" plush platypus	$22.95